PARAMETRIC ADDITIVE NUMBER THEORY

José Alfonso López Nicolás

Thanks to God, who is my Life, my Glory, my Honour, my saving Rock; to Inma, my beloved wife; and to all my family for their inconditional support.

Contents

Chapter 1

Introduction

In this original monograph we present sufficient conditions to know if given a positive real number $x = 2s$ and a uniformly discrete set of positive real numbers $\mathcal{P}$, such a number is close enough to the sum of two elements of this set when $s \notin \mathcal{P}$. In fact these conditions allows us to obtain bounds for the distance between $x = 2s$ and the sum set $\mathcal{P} + \mathcal{P}$ in a constructive way. The properties of the distribution function of $\mathcal{P}$, π, play an essential role, as expected. We study the case consist that the distribution function of $\mathcal{P}$ is subadditive, what is connected to the set of the prime numbers through the Second Hardy-Littlewood Conjecture. Moreover, we also study the wider case consisting of π being relatively subadditive, what is verified by the distribution function of prime numbers (proved by Pierre Dusart in his doctoral thesis). We also distinguish the cases when the distribution function is relatively contractive and when it is not, and obtain results in both cases, using the distribution function in order to estimate distances to $\mathcal{P} + \mathcal{P}$.

Given a positive real number $x = 2s$ and a uniformly discrete set of positive real numbers $\mathcal{P}$, we investigate sufficient conditions to determine bounds of the distance between such a number and the sums of two elements of this set. We obtain several constructive results which allows us to know these bounds and approximations to the sum set $\mathcal{P} + \mathcal{P}$. First we obtain an upper approximation of $x = 2s$ by a sum of two numbers of $\mathcal{P}$, and then we adjust and improve this approximation using the properties of the distribution function consisting of being absolutely or relatively subadditive and also using the eccentricity of the half of the given number, and distinguising between distribution functions which are relatively contractive and those which are not. From other point of view we obtain estimates for the elements of $\mathcal{P} + \mathcal{P}$. We use upper bounds for the distribution function in order to obtain results of approximation. We also study these results in the context of the prime numbers.

In Chapter 9 we obtain new conditions to determine if given a real number and a uniformly discrete set of real numbers, such a number can be expressed as sum of two elements of this set. Although we obtain several general results, our work is motivated by the particular case of the prime numbers set. Namely, we establish a relationship between bounds for the distribution function such as that of the Second Hardy-Littlewood Conjecture and the Goldbach's Property for uniformly discrete sequences of real numbers.

Parametric Number Theory is the part of Number Theory which studies the distribution of uniformly discrete sets of real numbers when their distribution function is completely known except for one or several real parameters, what allows us to obtain some information on the distribution of such sets, and, of course, results on the distribution of uniformly discrete sets whose distribution functions belong to a same *family*.

1.1 Note about the author

The author is Graduated in Mathematics by University of Murcia and has a Master in Advanced Mathematics by the same university. He is a researcher in Mathematical Analysis, namely in Functional Analysis and Number Theory. Besides he has cooperated with the Group of Research in Functional Analysis at University of Murcia and has published a celebrated article on additive number theory entitled *Advances in Additive Number Theory* (see [12]) along with other ones in Functional Analysis, and several books.

1.2 Notations

We establish some notation. We denote by $\mathbb{Z}^+ := \{m \in \mathbb{Z} \mid m > 0\}$ the set of positive integer numbers and $\mathbb{Z}^- := \{m \in \mathbb{Z} \mid m < 0\}$ the set of negative integer numbers. Given $A \subseteq \mathbb{R}^n$, we define $xA := \{a\,x \mid a \in A\}$, $A + B := \{a + b \mid a \in A,\, b \in B\}$ for every $B \subseteq \mathbb{R}^n$. We denote by $\mathbb{P}$ the set of prime numbers, and $\mathbb{P}^* := \mathbb{P} \setminus \{2\}$. We denote the number of elements of a set A by $Card(A)$. Given $x \in \mathbb{R}$ we denote by $[x]$ its integer part, and by $\lceil x \rceil$ the minimum integer greater or equal than x. We also denote the set of positive real numbers by $\mathbb{R}^+$. As usual, when we write $x > t$ (respectively, $x < t$) for a given real number t, without specifying any set which x belongs to, we mean that $x \in \mathbb{R}$, $x > t$ (respectively, $x \in \mathbb{R}$, $x < t$), and the same for whichever variable instead of x.

1.3   Structure of the book

This book is structured as follows. Chapters 1, 2 and 3 contain the introduction, our definitions and the main results, respectively. In Chapter 4 we study some properties of the distribution function of the prime numbers set, which justify our hypotheses. In Chapter 5 we prove Theorem 3.0.10. In Chapter 6 we prove Theorem 3.0.11. Chapter 7 is devoted to obtaining analogous results to the main ones without the condition over the distribution function of being relatively contractive. In Chapter 8 we study the approximation $p_{\pi(s)} + p_{\pi(s)+1} \approx 2s$ for subadditive distribution functions.

Finally, in Chapter 9 we establish a relationship between certain type of bounds for the distribution function such as that of the Second Hardy-Littlewood Conjecture and the Goldbach's Property for uniformly discrete sequences of real numbers.

Chapter 2

Definitions

We work with the concepts of distribution function cf a set of positive real numbers $\mathcal{P}$. We also use other concepts which help us to estimate the distance between a given positive real number and the set $\mathcal{P} + \mathcal{P}$, such as subadditive distribution function, relatively contractive distributicn function and eccentricity of a real number respect to a set of real numbers.

Definition 2.0.1. *Let $\mathcal{P} = (p_i)_{i \in \mathbb{Z}^+}$ be an strictly increasing sequence in $\mathbb{R}^+$.*

1. *We say that $\mathcal{P}$ is uniformly discrete (briefly, u.d.) if*

$$\inf_{i \in \mathbb{Z}^+} \{p_{i+1} - p_i\} > 0.$$

2. *The function $\pi := \pi_{\mathcal{P}} : [0, +\infty) \to \mathbb{R}$ defined by*

$$\pi(x) := Card\left(\{p \in \mathcal{P} \mid p \leq x\}\right) = Card\left(\mathcal{P} \cap (0, \, x]\right),$$

for each $x \in [0, +\infty)$, is called the distribution function of $\mathcal{P}$. π is monotonically increasing.

3. *$p_x := p_{[x]}$ for every $x \in [1, +\infty)$.*

Remark 2.0.2. *Notice these consequences of the definition of π:*

1. *For every $x \in \mathbb{R}^+$ we have $\{p_1, ..., p_{\pi(x)}\} \subseteq (0, \, x]$, and then $p_{\pi(x)} \leq x$.*

2. *$\left(p \leq x \Leftrightarrow p \leq p_{\pi(x)}\right)$ for every $x \in \mathbb{R}^+$, $p \in \mathcal{P}$.*

Definition 2.0.3. *Let $\mathcal{P} = (p_i)_{i \in \mathbb{Z}^+}$ be an strictly increasing and u.d. sequence in $\mathbb{R}^+$, and let $\pi := \pi_{\mathcal{P}} : [0, +\infty) \to \mathbb{R}$ be its distribution function.*

1. *We say that π is subadditive if*

$$\pi(x + y) \leq \pi(x) + \pi(y) \text{ for all } x, y \in (0, +\infty).$$

2. *We say that π is relatively subadditive if there exist constants $d, D > 0$, $d < D$, such that*

$$\pi(x + y) \leq \pi(x) + \pi(y) \text{ for all } x, y > 0, \, d < \frac{y}{x} < D.$$

3. *We say that π is $\mathcal{P}$-subadditive if $\pi(x+y) \leq \pi(x)+\pi(y)$ for all $x, y \in \mathcal{P}$. That is, if*

$$\pi(p_i + p_j) \leq i + j \text{ for every } i, j \in \mathbb{Z}^+.$$

Remark 2.0.4. *In the context of prime numbers, Second Hardy-Littlewood Conjecture precisely states that the distribution function of the prime numbers set is subadditive (see [9] and [7]).*

Definition 2.0.5. *Let $\mathcal{P} = (p_i)_{i \in \mathbb{Z}^+}$ be an strictly increasing and u.d. sequence in $\mathbb{R}^+$, and let $\pi := \pi_{\mathcal{P}} : [0, +\infty) \to \mathbb{R}$ be its distribution function.*

1. *We say that π is relatively contractive if for every $t > 0$ there exists a constant $D_t > 0$ such that*

$$\pi(x - y) \leq D_t \left(\pi(x) - \pi(y) \right)$$

for all $x, y > 0$ verifying that $y \leq x \leq t$.

2. *We say that π is contractive if there exists a constant $D > 0$ such that*

$$\pi(x - y) \leq D \left(\pi(x) - \pi(y) \right)$$

for all $x, y > 0$, $y \leq x$.

Remark 2.0.6. *Let $\mathcal{P} = (p_i)_{i \in \mathbb{Z}^+}$ be an strictly increasing and u.d. sequence in $\mathbb{R}^+$, and let $\pi : [0, +\infty) \to \mathbb{R}$ be its distribution function.*

1. π is relatively contractive if and only if for every $t > 0$ there exists a constant $E_t > 0$ such that

$$\pi(y) + E_t \, \pi(z) \le \pi(y + z)$$

for all y, $z \in \mathbb{R}$ verifying that $y > 0$, $z \ge 0$, $y + z \le t$.

2. π is contractive if and only if there exists a constant $E > 0$ such that

$$\pi(y) + E \, \pi(z) \le \pi(y + z)$$

for all y, $z \in \mathbb{R}$, $y > 0$, $z \ge 0$.

Remark 2.0.7. Let $\mathcal{P} = (p_i)_{i \in \mathbb{Z}^+}$ be an strictly increasing and u.d. sequence in $\mathbb{R}^+$, and let $\pi : [0, +\infty) \to \mathbb{R}$ be its distribution function. The following statements are equivalent:

1. π is relatively contractive.

2. $\pi(x) = \pi(y) \Rightarrow \pi(x - y) = 0$ for all x, $y > 0$, $x \ge y$.

3. $x - y \ge p_1 \Rightarrow \mathcal{P} \cap (y, \, x] \ne \emptyset$ for all x, $y > 0$, $x \ge y$.

4. $\mathcal{P} \cap (t, \, t + p_1] \ne \emptyset$ for all $t > 0$.

5. $p_{i+1} - p_i \le p_1$ for every $i \in \mathbb{Z}^+$.

Remark 2.0.7 determine the sequence of positive numbers with relative contractive distribution function and give us lots of examples of such sequences, as $\mathcal{P} = \mathbb{Z}^+$.

An important example of sequence of positive numbers with distribution function which is not relative contractive is the following one.

Example 2.0.8. Given $m \in \mathbb{Z}$, $m \ge 2$, consider the sequence of the m powers of the positive integer numbers: $\mathcal{P}(m) = (p_i := i^m)_{i \in \mathbb{Z}^+}$. Its distribution function π is given by

$$\pi(x) = \left[x^{1/m} \right] \quad \text{for each } x \in \mathbb{R}^+.$$

We can easily prove that

$$\pi(x - y) \le \left(\sum_{k=1}^{m} \frac{m!}{k! \, (m - k)!} \, \pi(x)^k - \pi(y)^m \right)^{1/m} \quad \text{for all } x, y \in \mathbb{R}, \, x \ge y.$$

π is not relative contractive (and consequently nor contractive).

Next we establish our general setting.

Let $\mathcal{P} = (p_i)_{i \in \mathbb{Z}^+}$ be an strictly increasing and u.d. sequence in $\mathbb{R}^+$. Let $s > 0$ be verifying that $s \geq p_1$. We wonder if $2s \in \mathcal{P} + \mathcal{P}$ (the optimal case). If $s \in \mathcal{P}$ (the trivial case), then $2s = s + s \in \mathcal{P} + \mathcal{P}$ and the answer is affirmative. We want to know what happens if $s \notin \mathcal{P}$. Assume that $2s \in \mathcal{P} + \mathcal{P}$, with $s \notin \mathcal{P}$. There exist $p, q \in \mathcal{P}$, with $p \leq q$, such that $2s = p + q$. Then $2s \geq 2p$ and, therefore $p \leq s$, or what is equivalent, $p \leq p_{\pi(s)}$. Thus we have that $p \in \mathcal{P}$, $2s - p \in \mathcal{P}$ and $p \leq p_{\pi(s)}$. So these last three conditions altogether are equivalent to $2s \in \mathcal{P} + \mathcal{P}$ when $s \notin \mathcal{P}$. Hence, if $2s = p_k + p_l$ with $1 \leq k < l$ and $s \notin \mathcal{P}$, then $k \leq \pi(s)$ and $\pi(s) + 1 \leq l \leq \pi(2s - p_1)$.

Definition 2.0.9. *Let* $\mathcal{P} = (p_i)_{i \in \mathbb{Z}^+}$ *be an strictly increasing and u.d. sequence in* $\mathbb{R}^+$. *Let* $s \in \mathbb{R}$ *be such that* $s \geq p_1$.

1. *Let* $k_s, l_s \in \mathbb{Z}^+$ *be such that* $k_s \leq \pi(s)$ *and* $\pi(s) + 1 \leq l_s \leq \pi(2s - p_1)$. *We define:*

   - $a_s := \pi(s) - k_s \in \{0, ..., \pi(s) - 1\}$.
   - $b_s := l_s - \pi(s) \in \{1, ..., \pi(2s - p_1) - \pi(s)\}$.

   Then $k_s = \pi(s) - a_s$, $l_s = \pi(s) + b_s$. *Observe that* $k_s \leq \pi(s) < l_s$. *Then* $k_s < l_s$, *and consequently* $p_{k_s} < p_{l_s}$.

2. *We call eccentricity of* s *with respect to* $\mathcal{P}$ *to the integer number*

$$e_s := 2\,\pi(s) - \pi(2s).$$

   Observe that if π *is subadditive, then* $e_s \geq 0$.

Notice that $p_{k_s} \leq s$ and $p_{l_s} \leq 2s - p_1$.

We want to obtain bounds for the distance between $2s$ and $p_{k_s} + p_{l_s}$.

Chapter 3

Main results

Our most important results until Chapter 8 (but not the unique ones) are the following theorems (Chapter 9 is an article by its own with its main results, especially on Golbach Conjecture).

Theorem 3.0.10. *Let $\mathcal{P} = (p_i)_{i \in \mathbb{Z}^+}$ be an strictly increasing and u.d. sequence in $\mathbb{R}^+$. Assume that the distribution function π is relatively contractive. Let $s \in \mathbb{R}$ be such that $s \geq p_1$, $s \notin \mathcal{P}$. Let $a > 0$. Suppose that there exists $m_a \in \mathbb{Z}^+$ such that*

1. *There exists a constant $B_a > 0$ verifying that*

$$p_{(a \cdot m)} \geq B_a \cdot p_m \text{ for every } m \in \mathbb{Z}, \ m \geq m_a.$$

2. *There exists a constant $C_a > 0$ such that*

$$\pi(a \cdot x) \geq C_a \cdot \pi(x) \text{ for every } x \geq p_{m_a}.$$

3. *$\pi(x + y) \leq \pi(x) + \pi(y)$ for all $x, y \in [p_{m_c}, +\infty)$.*

Let $a_s \in \{0, ..., \pi(s) - 1\}$ and $b_s \in \{1, ..., \pi(2s - p_1) - \pi(s)\}$. Assume that the following conditions are verified:

a) *$b_s \geq (a - 1)\pi(s) - a \cdot a_s$.*

b) *$a_s \leq \pi(s) - m_a$.*

c) *$e_s = a_s - b_s$.*

11

d) $a_s < \pi(s) - \frac{\pi(2s)}{R}$, where $R := C_{(B_a+1)} > 0$ is the constant associated to $B_a + 1$ by hypothesis 2.

Then $2s < p_{k_s} + p_{l_s} < 2s + p_1$.

A key step in the proof of Theorem 3.0.10, as we will see in Section 2, is make sure that $\pi(p_{k_s} + p_{l_s}) \leq \pi(p_{k_s}) + \pi(p_{l_s})$. Hence we can replace the hypothesis 3 by the next one:

$$\pi(p_i + p_j) \leq i + j \text{ for every } i, j \in \mathbb{Z}^+, m_a \leq i < j.$$

When the distribution function π is relatively subadditive we may also obtain a constructive result, which is the following one.

Theorem 3.0.11. *Let $\mathcal{P} = (p_i)_{i \in \mathbb{Z}^+}$ be an strictly increasing and u.d. sequence in $\mathbb{R}^+$. Suppose that the distribution function π is relatively contractive. Let $s \in \mathbb{R}$, $s \geq p_1$, $s \notin \mathcal{P}$. Let $a > 0$. Assume that there exists $m_a \in \mathbb{Z}^+$ such that*

1. *There exists a constant $B_a > 0$ verifying that*

$$p_{(a \cdot m)} \geq B_a \cdot p_m \text{ for every } m \in \mathbb{Z}, m \geq m_a.$$

2. *There exists a constant $C_a > 0$ verifying that*

$$\pi(a \cdot x) \geq C_a \cdot \pi(x) \text{ for every } x \geq p_{m_a}.$$

Let $a_s \in \{0, ..., \pi(s) - 1\}$ and $b_s \in \{1, ..., \pi(2s - p_1) - \pi(s)\}$. Suppose that the following conditions are verified:

a) $b_s \geq (a-1)\pi(s) - a \cdot a_s$.

b) $a_s \leq \pi(s) - m_a$.

c) $e_s = a_s - b_s$.

d) $a_s < \pi(s) - \frac{\pi(2s)}{R}$, with $R := C_{(B_a+1)} > 0$ the constant associated to $B_a + 1$ by hypothesis 2.

Then:

i) Suppose that there exists a constant $d > 0$ such that

$$\pi(x + y) \leq \pi(x) + \pi(y) \text{ for all } x, \ y > 0, \ d \leq \frac{y}{x}.$$

Let $r_d > 0$ be such that $p_{(r_d \cdot k_s)} \geq d \cdot p_{k_s}$. If $r_d \left(\pi(s) - a_s\right) < \pi(s) + b_s$, then $2s < p_{k_s} + p_{l_s} < 2s + p_1$.

ii) Assume that there exists a constant $E > 0$ such that

$$\pi(x + y) \leq \pi(x) + \pi(y) \text{ for all } x, \ y > 0, \ E \geq \frac{y}{x}.$$

If $\pi(s) + b_s < C_E \left(\pi(s) - a_s\right)$, where $C_E > 0$ is the constant associated to E as in hypothesis 2, then $2s < p_{k_s} + p_{l_s} < 2s + p_1$. If $E < 1$, we add the assumption consisting of $E \cdot p_{k_s} \geq p_1$.

iii) Suppose that there exist constants $0 < d \leq E$ such that

$$\pi(x + y) \leq \pi(x) + \pi(y) \text{ for all } x, \ y > 0, \ d \leq \frac{y}{x} \leq E.$$

Let $r_d > 0$ be such that $p_{(r_d \cdot k_s)} \geq d \cdot p_{k_s}$.

If $r_d \left(\pi(s) - a_s\right) < \pi(s) + b_s < C_E \left(\pi(s) - a_s\right)$, then $2s < p_{k_s} + p_{l_s} < 2s + p_1$.

If $E < 1$, we add the assumption consisting of $E \cdot p_{k_s} \geq p_1$.

Chapter 4

Motivation: the prime number set

In this chapter we study our hypotheses in relationship with the prime number set

$$\mathbb{P} = \{p_1 = 2, p_2 = 3, p_3 = 5, p_4 = 7, \ldots\}.$$

We obtain the following results from [4]:

Remark 4.0.12.

i) $p_{a \cdot b} > a\, p_b$ for all a, $b \in \mathbb{Z}$, a, $b \geq 2$ (if $b = 1$ this inequality is true for every $a \geq 5$). (See Proposition 1.11, p. 44).

ii) $\pi(x + y) \leq \pi(x) + \pi(y)$ for all x, $y \in \mathbb{R}$, $1 \leq x \leq y \leq \frac{7}{5} x \log x \log_2 x$. (See Theorem 2.6, p. 73).

iii) $\pi(x + y) \leq \pi(x) + \pi(y)$ for all x, $y \in [3, +\infty)$ such that $\frac{1}{109} \leq \frac{y}{x} \leq 1$. (See Theorem 2.5, p. 67).

iv) $\frac{x}{\log x - 1} \leq \pi(x) \leq \frac{x}{\log x - 1.1}$ for all $x \in [60184, +\infty)$, where the first inequality is true even for every $x \in [5393, +\infty)$. (See Theorem 1.10, p. 37, and Theorem 6.9 of [5]). From these inequalities we obtain that

$$\pi(a\,x) \geq a\, \frac{1 - \frac{1.1}{\log 60184}}{1 + \alpha} \pi(x)$$

14

for all $\alpha \in \mathbb{R}^+$, $a \in [1, 60184^\alpha]$, $x \in [60184, +\infty)$; and

$$\pi(a\,x) \le a\,\frac{\log(60184) - 1}{\log(60184) - 1.1}\,\pi(x) \le 1.010095746224\,a\,\pi(x)$$

for all $a \in [1, +\infty)$, $x \in [60184, +\infty)$.

Rosser and Schoendfeld proved that $\pi(2x) \le 2\,\tau(x)$ for all $x \in [3, +\infty)$ and therefore the eccentricity $e_x = 2\,\pi(x) - \pi(2x)$ is non negative for these numbers. Moreover, they proved that $\pi(2x) < 2\,\pi(x)$ for $x \ge 11$ (see [13], and also [11]), so $e_x > 0$ for each $x \ge 11$.

Examples 4.0.13. *We have:*

i) $\pi(2x) \ge 1.693424\,\pi(x)$ for all $x \in [60184, +\infty)$ (taking $\alpha = 0.06299$), so $C_2 = 1.693424$.

ii) $\pi(3x) \ge 2.45505\,\pi(x)$ for all $x \in [60184, +\infty)$ (taking $\alpha = 0.09983$), and so $C_3 = 2.45505$.

iii) $\pi(109x) \ge 68.78357037\,\pi(x)$ for every $x \in [60184, +\infty)$ (taking $\alpha = 0.426287$), thus $C_{109} = 68.78357037$.

Remark 4.0.14. *From [5] we have the inequalities*

$$\frac{x}{\log x}\left(1 + \frac{1}{\log x} + \frac{2}{\log^2 x}\right) \le \pi(x) \le \frac{x}{\log x}\left(1 + \frac{1}{\log x} + \frac{2.334}{\log^2 x}\right),$$

where the first inequality is verified for all $x \ge 88783$ and the second one is for all $x \ge 2953652287$. From here we obtain that

1.

$$e_s = 2\,\pi(s) - \pi(2s) \le$$

$$\le 2s\left(\frac{\log 2}{(\log s)(\log 2 + \log s)} + \frac{\log^2 2 + 2\log 2\log s}{(\log s)^2(\log 2 + \log s)^2}\right) +$$

$$+4s\left(\frac{1.167}{\log^3 s} - \frac{1}{(\log 2 + \log s)^3}\right),$$

2.

$$e_s \geq 2s \left(\frac{\log 2}{(\log s)(\log 2 + \log s)} + \frac{\log^2 2 + 2\log 2 \log s}{(\log s)^2 (\log 2 + \log s)^2} \right) -$$

$$-4s \left(\frac{1.167}{(\log 2 + \log s)^3} - \frac{1}{\log^3 s} \right),$$

for every $s \geq 2953652287$.

Chapter 5

Proof of Theorem 3.0.10

In this chapter we will prove Theorem 3.0.10. First we need the following auxiliary result.

Lemma 5.0.15. *Let $\mathcal{P} = (p_i)_{i \in \mathbb{Z}^+}$ be an strictly increasing and u.d. sequence in $\mathbb{R}^+$. Let $s \in \mathbb{R}$ be such that $s \geq p_1$, $s \notin \mathcal{P}$. Let $a > 0$. Suppose that there exists $m_a \in \mathbb{Z}^+$ such that*

1. There exists a constant $B_a > 0$ such that

$$p_{(a \cdot m)} \geq B_a \cdot p_m \text{ for every } m \in \mathbb{Z}^+, \, m \geq m_a.$$

2. There exists a constant $C_a > 0$ verifying that

$$\pi(a \cdot x) \geq C_a \cdot \pi(x) \text{ for every } x \geq p_{m_a}.$$

Let $a_s \in \{0, ..., \pi(s) - 1\}$ and $b_s \in \{1, ..., \pi(2s - p_1) - \pi(s)\}$ be such that

$$b_s \geq (a - 1)\pi(s) - a \cdot a_s.$$

Also assume that $a_s \leq \pi(s) - m_a$ and $a_s < \pi(s) - \frac{\pi(2s)}{R}$, where $R := C_{(B_a+1)} > 0$ is the constant associated to $B_a + 1$ by hypothesis 2. Then $2s < p_{k_s} + p_{l_s}$.

Proof. We will prove that $\pi(p_{k_s} + p_{l_s}) > \pi(2s)$, and from this we will obtain that $2s < p_{k_s} + p_{l_s}$.

As $a_s \leq \pi(s) - m_a$ then $k_s \geq m_a$. We also have that $l_s \geq a \cdot k_s$ because $b_s \geq (a - 1)\pi(s) - a \cdot a_s$. Hence

$$p_{l_s} \geq p_{a \cdot k_s} \geq B_a \cdot p_{k_s}.$$

Then $p_{l_s} \geq B_a \cdot p_{k_s}$. Therefore

$$\pi \left(p_{k_s} + p_{l_s}\right) \geq \pi \left(\left(B_a + 1\right) p_{k_s}\right) \geq C_{(B_a+1)} \pi(p_{k_s}) =$$
$$= R\, k_s = R\, \left(\pi(s) - a_s\right).$$

If $R\,\left(\pi(s) - a_s\right) > \pi(2s)$, we will obtain that $\pi\left(p_{k_s} + p_{l_s}\right) > \pi(2s)$, and then $2s < p_{k_s} + p_{l_s}$.

The following statements are equivalent:

i) $R\,\left(\pi(s) - a_s\right) > \pi(2s)$.

ii) $R \cdot a_s < R\,\pi(s) - \pi(2s)$.

iii) $a_s < \pi(s) - \frac{\pi(2s)}{R}$.

Hence we have that $R\,\left(\pi(s) - a_s\right) > \pi(2s)$, what concludes the proof. $\square$

Remark 5.0.16. *Notice that* $\pi(s) - \frac{\pi(2s)}{R} = \frac{(R-2)\pi(s)+e_s}{R}$*, where* e_s *is the eccentricity of* s*.*

Proof of Theorem 3.0.10. By Lemma 5.0.15 we have $2s < p_{k_s} + p_{l_s}$. As $p_{k_s} + p_{l_s} \leq 3s$ and π is relatively contractive, then there exists a constant $D_{3s} > 0$ such that

$$\pi(x - y) \leq D_{3s}\,\left(\pi(x) - \pi(y)\right)$$

for all $x,\, y \in \mathbb{R}^+$ verifying that $y \leq x \leq 3s$. Then:

$$0 \leq \pi\left(\left(p_{k_s} + p_{l_s}\right) - 2s\right) \leq D_{3s}\,\left(\pi\left(p_{k_s} + p_{l_s}\right) - \pi(2s)\right) \leq$$
$$\leq D_{3s}\,\left(\pi\left(p_{k_s}\right) + \pi\left(p_{l_s}\right) - \pi(2s)\right) = D_{3s}\,\left(k_s + l_s - \pi(2s)\right) =$$
$$= D_{3s}\,\left(\pi(s) - a_s + \pi(s) + b_s - \pi(2s)\right) = D_{3s}\,\left(e_s - \left(a_s - b_s\right)\right) = 0.$$

Therefore $\pi\left(\left(p_{k_s} + p_{l_s}\right) - 2s\right) = 0$, which is equivalent to

$$\left(p_{k_s} + p_{l_s}\right) - 2s < p_1,$$

and so $2s < p_{k_s} + p_{l_s} < 2s + p_1$. $\square$

A slight generalization of Theorem 3.0.10 is the following result.

Theorem 5.0.17. *Let* $\mathcal{P} = (p_i)_{i\in\mathbb{Z}^+}$ *be an strictly increasing and u.d. sequence in* $\mathbb{R}^+$*. Suppose that the distribution function* π *is relatively contractive. Let* $s \in \mathbb{R}$ *be such that* $s \geq p_1$*,* $s \notin \mathcal{P}$*. Let* $a > 0$*. Assume that there exists* $m_a \in \mathbb{Z}^+$ *verifying that*

1. *There exists $B_a > 0$ verifying that*

$$p_{(a \cdot m)} \geq B_a \cdot p_m \text{ for every } m \in \mathbb{Z}, \ m \geq m_a.$$

2. *There exists $C_a > 0$ such that*

$$\pi (a \cdot x) \geq C_a \cdot \pi(x) \text{ for every } x \geq p_{m_a}.$$

3. *There exists $D_a \in \mathbb{R}$ such that*

$$\pi(x + y) \leq \pi(x) + \pi(y) + D_a \text{ for all } x, \ y \in [p_{m_a}, +\infty).$$

Let $a_s \in \{0, ..., \pi(s) - 1\}$ and $b_s \in \{1, ..., \pi(2s - p_1) - \pi(s)\}$. Assume that the following conditions are verified:

a) $b_s \geq (a - 1)\pi(s) - a \cdot a_s$.

b) $a_s \leq \pi(s) - m_a$.

c) $e_s + D_a = a_s - b_s$.

d) $a_s < \pi(s) - \frac{\pi(2s)}{R}$, with $R := C_{(B_a + 1)} > 0$.

Then $2s < p_{k_s} + p_{l_s} < 2s + p_1$.

Proof. Again by Lemma 5.0.15, $2s < p_{k_s} + p_{l_s}$. As $p_{k_s} + p_{l_s} \leq 3s$ and π is relatively contractive, then there exists a constant $D_{3s} > 0$ such that

$$\pi(x - y) \leq D_{3s} \left(\pi(x) - \pi(y) \right)$$

for every $x, \ y \in \mathbb{R}^+$ such that $y \leq x \leq 3s$. Thus:

$$0 \leq \pi \left((p_{k_s} + p_{l_s}) - 2s \right) \leq D_{3s} \left(\pi \left(p_{k_s} + p_{l_s} \right) - \pi(2s) \right) \leq$$
$$\leq D_{3s} \left(\pi \left(p_{k_s} \right) + \pi \left(p_{l_s} \right) + D_a - \pi(2s) \right) = D_{3s} \left(k_s + l_s + D_a - \pi(2s) \right) =$$
$$= D_{3s} \left(\pi(s) - a_s + \pi(s) + b_s + D_a - \pi(2s) \right) =$$
$$= D_{3s} \left(e_s + D_a - (a_s - b_s) \right) = 0.$$

Hence $\pi \left((p_{k_s} + p_{l_s}) - 2s \right) = 0$, that is equivalent to

$$(p_{k_s} + p_{l_s}) - 2s < p_1,$$

and then $2s < p_{k_s} + p_{l_s} < 2s + p_1$. $\qquad \square$

Chapter 6

Proof of Theorem 3.0.11

In this chapter we will prove Theorem 3.0.11.

Proof. This proof is essentially the same as the proof of Theorem 3.0.10. We only have to prove that $\pi\left(p_{k_s} + p_{l_s}\right) \leq \pi\left(p_{k_s}\right) + \pi\left(p_{l_s}\right)$.

  i) We will prove that $d \leq \frac{p_{l_s}}{p_{k_s}}$ and then we will obtain

$$\pi\left(p_{k_s} + p_{l_s}\right) \leq \pi\left(p_{k_s}\right) + \pi\left(p_{l_s}\right).$$

As $p_{(r_d \cdot k_s)} \geq d \cdot p_{k_s}$, then $\pi\left(p_{(r_d \cdot k_s)}\right) \geq \pi\left(d \cdot p_{k_s}\right)$. Our assumption

$$r_d\left(\pi(s) - a_s\right) < \pi(s) + b_s$$

is equivalent to $r_d \cdot k_s < l_s$. Then

$$\pi\left(p_{(r_d \cdot k_s)}\right) < \pi\left(p_{l_s}\right).$$

Thus $\pi\left(d\, p_{k_s}\right) < \pi\left(p_{l_s}\right)$, what implies that $d \cdot p_{k_s} \leq p_{l_s}$.

  ii) We will prove that $\frac{p_{l_s}}{p_{k_s}} \leq E$ and then we will obtain

$$\pi\left(p_{k_s} + p_{l_s}\right) \leq \pi\left(p_{k_s}\right) + \pi\left(p_{l_s}\right).$$

If $E < 1$, then by hypothesis $E \cdot p_{k_s} \geq p_1$ and then $\pi\left(E \cdot p_{k_s}\right) \geq 1$, and we define

$$C_E := \frac{\pi\left(E \cdot p_{k_s}\right)}{\pi\left(p_{k_s}\right)}.$$

Therefore, whatever value of $E > 0$ we have that $\pi\left(E \cdot p_{k_s}\right) \geq C_E\, \pi\left(p_{k_s}\right)$, by hypothesis 2.

The assumption $\pi(s) + b_s < C_E\left(\pi(s) - a_s\right)$ is equivalent to $l_s < C_E \cdot k_s$, which is also equivalent to $\pi\left(p_{l_s}\right) < C_E\, \pi\left(p_{k_s}\right)$. Thus $\pi\left(p_{l_s}\right) < \pi\left(E \cdot p_{k_s}\right)$, and then $p_{l_s} \leq E\, p_{k_s}$.

iii) This is an immediate consequence of the two previous items.

$\square$

Chapter 7

Results for distribution function not necessarily relatively contractive

In this chapter we obtain results which allow us to estimate

$$\pi\left((p_{k_s} + p_{l_s}) - 2s\right),$$

and consequently also

$$(p_{k_s} + p_{l_s}) - 2s,$$

when the distribution function of $\mathcal{P}$ is not necessarily relatively contractive but verifies certain different condition of upper boundness depending on one or more parameters (this is called subparametric distribution).

Indeed, notice that given $x, C > 0$, we have that $\pi(x) \leq C$ is equivalent to $x < p_{([C]+1)}$.

Theorem 7.0.18. *Let $\mathcal{P} = (p_i)_{i \in \mathbb{Z}^+}$ be an strictly increasing and u.d. sequence in $\mathbb{R}^+$ with $p_1 > 1$. Let $s \in \mathbb{R}$ such that $s \geq p_1$, $s \notin \mathcal{P}$.*

Let $a \in \mathbb{R}$, $a > 0$. Suppose that there exists $m_a \in \mathbb{Z}^+$ such that

1. *There exists $B_a > 0$ such that that*

$$p_{(a \cdot m)} \geq B_a \cdot p_m \text{ for every } m \in \mathbb{Z}, \ m \geq m_a.$$

2. *There exists $C_a > 0$ verifying that*

$$\pi\left(a \cdot x\right) \geq C_a \cdot \pi(x) \text{ for every } x \geq p_{m_a}.$$

3. $\pi(x + y) \leq \pi(x) + \pi(y)$ for all x, $y \in [p_{m_a}, +\infty)$.

Let $a_s \in \{0, ..., \pi(s) - 1\}$ and $b_s \in \{1, ..., \pi(2s - p_1) - \pi(s)\}$ be such that

$$b_s \geq (a - 1)\pi(s) - a \cdot a_s.$$

Assume that the following three conditions are verified:
  $a_s \leq \pi(s) - m_a$, $a_s - b_s = e_s$ and $a_s < \pi(s) - \frac{\pi(2s)}{R}$, with $R := C_{(B_a+1)} > 0$
the constant associated to $B_a + 1$ by hypothesis 2.
  Also assume that there exists a constant $C > 0$ such that

$$\pi\left(\frac{x}{y}\right) \leq C\ (\pi(x) - \pi(y))\ \text{for all}\ x, y \in \mathbb{R},\ 2\,p_1 \leq y \leq x.$$

Then

$$2s < p_{k_s} + p_{l_s} < 2s \cdot p_1.$$

Proof. By Lemma 5.0.15 we have $2s < p_{k_s} + p_{l_s}$.
  Then we have:

$$0 \leq \pi\left(\frac{p_{k_s} + p_{l_s}}{2s}\right) \leq C\ (\pi\,(p_{k_s} + p_{l_s}) - \pi(2s)) \leq$$
$$\leq C\ (\pi\,(p_{k_s}) + \pi\,(p_{l_s}) - \pi(2s)) =$$
$$= C\ (k_s + l_s - \pi(2s)) = C\ (\pi(s) - a_s + \pi(s) + b_s - \pi(2s)) =$$
$$= C\ (e_s - (a_s - b_s)) = 0.$$

Therefore $\pi\left(\frac{p_{k_s} + p_{l_s}}{2s}\right) = 0$, which is equivalent to

$$\frac{p_{k_s} + p_{l_s}}{2s} < p_1.$$

Hence $2s < p_{k_s} + p_{l_s} < 2s \cdot p_1$. □

Observe that as in Theorem 3.0.10 we can replace the hypothesis 3 by the next one:

$$\pi\,(p_i + p_j) \leq i + j\ \text{for every}\ i, j \in \mathbb{Z}^+,\ m_a \leq i < j.$$

Relaxing the condition of subadditivity for the distribution function we have the following result.

Theorem 7.0.19. *Let $\mathcal{P} = (p_i)_{i \in \mathbb{Z}^+}$ be an strictly increasing and u.d. sequence in $\mathbb{R}^+$ with $p_1 > 1$. Let $s \in \mathbb{R}$, $s \geq p_1$, $s \notin \mathcal{P}$. Let $a > 0$. Suppose that there exists $m_a \in \mathbb{Z}^+$ such that*

1. *There exists a constant $B_a > 0$ verifying that*

$$p_{(a \cdot m)} \geq B_a \cdot p_m \text{ for every } m \in \mathbb{Z}, \ m \geq m_a.$$

2. *There exists a constant $C_a > 0$ verifying that*

$$\pi(a \cdot x) \geq C_a \cdot \pi(x) \text{ for every } x \geq p_{m_a}.$$

Let $a_s \in \{0, ..., \pi(s) - 1\}$ and $b_s \in \{1, ..., \pi(2s - p_1) - \pi(s)\}$ be such that

$$b_s \geq (a - 1)\pi(s) - a \cdot a_s.$$

Assume that $a_s \leq \pi(s) - m_a$, $a_s - b_s = e_s$ and $a_s < \pi(s) - \frac{\pi(2s)}{R}$, where $R := C_{(B_a+1)} > 0$.

Also assume that there exists a constant $C > 0$ such that

$$\pi\left(\frac{x}{y}\right) \leq C \ (\pi(x) - \pi(y)) \text{ for all } \ x, y \in \mathbb{R}, \ 2 \, p_1 \leq y \leq x.$$

Then

i) Suppose that there exists a constant $d > 0$ such that

$$\pi(x + y) \leq \pi(x) + \pi(y) \text{ for all } x, y > 0, \ d \leq \frac{y}{x}.$$

Let $r_d > 0$ be such that $p_{(r_d \cdot k_s)} \geq d \cdot p_{k_s}$. If $r_d \ (\pi(s) - a_s) < \pi(s) + b_s$, then $2s < p_{k_s} + p_{l_s} < 2s \cdot p_1$.

ii) Assume that there exists a constant $E > 0$ such that

$$\pi(x + y) \leq \pi(x) + \pi(y) \text{ for all } x, y > 0, \ E \geq \frac{y}{x}.$$

If $\pi(s) + b_s < C_E \ (\pi(s) - a_s)$, where $C_E > 0$ is the constant associated to E as in hypothesis 2, then $2s < p_{k_s} + p_{l_s} < 2s \cdot p_1$. If $E < 1$, we add the assumption consisting of $E \cdot p_{k_s} \geq p_1$.

iii) Assume that there exist constants $0 < d \leq E$ such that

$$\pi(x + y) \leq \pi(x) + \pi(y) \text{ for all } x,\ y > 0,\ d \leq \frac{y}{x} \leq E.$$

Let $r_d > 0$ be such that $p_{(r_d \cdot k_s)} \geq d \cdot p_{k_s}$.

If $r_d\left(\pi(s) - a_s\right) < \pi(s) + b_s < C_E\left(\pi(s) - a_s\right)$, then $2s < p_{k_s} + p_{l_s} < 2s \cdot p_1$.
If $E < 1$, we add the hypothesis $E \cdot p_{k_s} \geq p_1$.

Proof. This proof is essentially the same as the proof of Theorem 3.0.11. We only have to prove that $\pi\left(p_{k_s} + p_{l_s}\right) \leq \pi\left(p_{k_s}\right) + \pi\left(p_{l_s}\right)$, and for this we do the same steps, line by line, as in the proof of Theorem 3.0.11 (see section 3). $\square$

Theorem 7.0.20. *Let $\mathcal{P} = (p_i)_{i \in \mathbb{Z}^+}$ be an strictly increasing and u.d. sequence in $\mathbb{R}^+$. Let $s \in \mathbb{R}$, $s \geq p_1$, $s \notin \mathcal{P}$. Let $a > 0$. Suppose that there exists $m_a \in \mathbb{Z}^+$ such that*

1. *There exists a constant $B_a > 0$ such that*

$$p_{(a \cdot m)} \geq B_a \cdot p_m \text{ for every } m \in \mathbb{Z},\ m \geq m_a.$$

2. *There exists a constant $C_a > 0$ verifying that*

$$\pi(a \cdot x) \geq C_a \cdot \pi(x) \text{ for every } x \geq p_{m_a}.$$

3. *$\pi(x + y) \leq \pi(x) + \pi(y)$ for all $x,\ y \in [p_{m_a}, +\infty)$.*

Let $a_s \in \{0, ..., \pi(s) - 1\}$ and $b_s \in \{1, ..., \pi(2s - p_1) - \pi(s)\}$ be such that

$$b_s \geq (a - 1)\pi(s) - a \cdot a_s.$$

Suppose that the following three conditions are also verified:
$a_s \leq \pi(s) - m_a$, $a_s - b_s - e_s = 0$ and $a_s < \pi(s) - \frac{\pi(2s)}{R}$, being $R := C_{(B_a + 1)} > 0$.

Let $\beta \in \mathbb{R}^+$. Assume that

i) $(x \geq y \Rightarrow x\,\pi(y) - y\,\pi(x) + \beta > 0)$ for all $x,\ y > 0$, $2\,p_1 \leq y$.

ii) There exists a constant $C > 0$ such that

$$\pi\left(\frac{x}{y}\right) \leq C\,\frac{\left(\frac{x}{y} - 1\right)\pi(x)\,\pi(y)}{x\,\pi(y) - y\,\pi(x) + \beta} \text{ for all } x,\ y \in \mathbb{R},\ 2\,p_1 \leq y \leq x.$$

Then

$$\pi\left(\frac{p_{k_s} + p_{l_s}}{2s}\right) \leq \frac{C}{\beta}\ (s - p_1)\ \left(\frac{\pi(2s)}{2s}\right)^2.$$

Proof. From Lemma 5.0.15 we obtain that $2s < p_{k_s} + p_{l_s}$. We also know that $p_{k_s} + p_{l_s} \leq 3s - p_1$.

$$\pi\left(\frac{p_{k_s} + p_{l_s}}{2s}\right) \leq C\ \frac{\left(\frac{p_{k_s}+p_{l_s}}{2s} - 1\right)\ \pi(p_{k_s} + p_{l_s})\ \pi(2s)}{(p_{k_s} + p_{l_s})\ \pi(2s) - 2s\ \pi(p_{k_s} + p_{l_s}) + \beta} \leq$$

$$\leq C\ \frac{\left(\frac{3s-p_1}{2s} - 1\right)\ (k_s + l_s)\ \pi(2s)}{2s\ (\pi(2s) - (k_s + l_s)) + \beta} =$$

$$= \frac{1}{2s}\ \frac{\frac{s-p_1}{2s}\ (\pi(s) - a_s + \pi(s) + b_s)\ \pi(2s)}{\pi(2s) - (\pi(s) - a_s + \pi(s) + b_s) + \beta} =$$

$$= \frac{C}{4s^2}\ \frac{(s - p_1)\ (2\,\pi(s) - a_s + b_s)\ \pi(2s)}{a_s - b_s - e_s + \beta} =$$

$$= \frac{C}{4s^2}\ \frac{(s - p_1)\ (e_s - a_s + b_s + \pi(2s))\ \pi(2s)}{a_s - b_s - e_s + \beta} =$$

$$= \frac{C}{4s^2}\ \frac{(s - p_1)\ (\pi(2s))^2}{\beta} = \frac{C}{\beta}\ (s - p_1)\ \left(\frac{\pi(2s)}{2s}\right)^2,$$

where in the first inequality we have used that

$$\pi\left(p_{k_s} + p_{l_s}\right) \leq \pi\left(p_{k_s}\right) + \pi\left(p_{l_s}\right) = k_s + l_s.$$

$\square$

Remark 7.0.21. *Theorem 7.0.20 is inspired by Number Prime Theorem (see [3]):*

$$\pi(t) \approx \frac{t}{\log t},\ t \to +\infty,$$

where π is the distribution function of the set of the prime numbers, $\mathbb{P}$.

For enough large x, $y \in \mathbb{R}^+$ such that $\frac{x}{y}$ is also enough large, we have:

1. *$\frac{x}{y} > \frac{\pi(x)}{\pi(y)}$.*

2. *$\pi\left(\frac{x}{y}\right) \approx \frac{\frac{x}{y}\ \pi(x)\ \pi(y)}{x\ \pi(y) - y\ \pi(x)}$.*

Obviously in this case we can also obtain a similar result to Theorem 7.0.19.

Now we obtain more results over estimates of $(p_{k_s} + p_{l_s}) - 2s$ using the value of $\pi\left((p_{k_s} + p_{l_s}) - 2s\right)$.

Theorem 7.0.22. *Let $\mathcal{P} = (p_i)_{i \in \mathbb{Z}^+}$ be an strictly increasing and u.d. sequence in $\mathbb{R}^+$.*

Let $a > 0$. Assume that there exists $m_a \in \mathbb{Z}^+$ such that

1. There exists a constant $B_a > 0$ verifying that

$$p_{(a \cdot m)} \geq B_a \cdot p_m \text{ for every } m \in \mathbb{Z}, \ m \geq m_a.$$

2. For every $t > 0$ there exist constants $C_t, \ D_t > 0, \ C_t \leq D_t$, verifying that

$$D_t \cdot \pi(x) \geq \pi(t \cdot x) \geq C_t \cdot \pi(x) \text{ for every } x \geq p_{m_a}.$$

3. $\pi(x + y) \leq \pi(x) + \pi(y)$ for all $x, \ y \geq p_{m_a}$.

Let $s \in \mathbb{R}$ be such that $s \geq p_{m_a}$, $s \notin \mathcal{P}$. We define $\alpha_s := \frac{3s - p_1}{p_1}$.

Let $a_s \in \{0, ..., \pi(s) - 1\}$ and $b_s \in \{1, ..., \pi(2s - p_1) - \pi(s)\}$ be such that

$$b_s \geq (a - 1)\pi(s) - a \cdot a_s.$$

Suppose that $a_s \leq \pi(s) - m_a$, $a_s - b_s = e_s$ and $a_s < \pi(s) - \frac{\pi(2s)}{R}$, where $R := C_{(B_a + 1)} > 0$.

i) Suppose that there exist constants $E > 0$, $F \geq 0$ and a sequence of positive numbers $(r_n)_{n \in \mathbb{Z}^+} \subseteq \mathbb{R}^+$ such that

$$\pi(x - y) \leq F + E \cdot \sum_{n=1}^{+\infty} \left(\pi\left(\frac{x}{r_n}\right) - \pi\left(\frac{y}{r_n}\right) \right) =$$

$$= F + E \cdot \sum_{n \in \mathbb{Z}^+, \ r_n \leq \frac{x}{p_1}} \left(\pi\left(\frac{x}{r_n}\right) - \pi\left(\frac{y}{r_n}\right) \right),$$

for all $x, \ y \in \mathbb{R}^+$, $2\,p_1 \leq y \leq x$. Then

$$\pi\left((p_{k_s} + p_{l_s}) - 2s\right) \leq F + E \cdot \pi(2s) \cdot \sum_{n \in \mathbb{Z}^+, \ r_n \leq \alpha_s} \left(D_{\frac{1}{r_n}} - C_{\frac{1}{r_n}} \right).$$

ii) Assume that there exist constants $E > 0$, $F \geq 0$ and a sequence of positive numbers $(r_n)_{n \in \mathbb{Z}^+} \subseteq \mathbb{R}^+$ verifying that

$$\pi(x - y) \leq F + E \cdot \frac{\sum_{n=1}^{+\infty} \left(\pi\left(\frac{x}{r_n}\right) - \pi\left(\frac{y}{r_n}\right) \right)}{\pi(y)} =$$

$$= F + E \cdot \frac{\sum_{n \in \mathbb{Z}^+, \, r_n \leq \frac{x}{p_1}} \left(\pi\left(\frac{x}{r_n}\right) - \pi\left(\frac{y}{r_n}\right) \right)}{\pi(y)},$$

for all x, $y \in \mathbb{R}^+$, $2\,p_1 \leq y \leq x$. Then

$$\pi\left((p_{k_s} + p_{l_s}) - 2s\right) \leq F + E \cdot \sum_{n \in \mathbb{Z}^+, \, r_n \leq \alpha_s} \left(D_{\frac{1}{r_n}} - C_{\frac{1}{r_n}} \right).$$

Proof. By Lemma 5.0.15 we have that $2s < p_{k_s} + p_{l_s}$. Besides we also know that $p_{k_s} + p_{l_s} \leq 3s - p_1$.

i)

$$\pi\left(p_{k_s} + p_{l_s} - 2s\right) \leq F + E \cdot \sum_{n \in \mathbb{Z}^+, \, r_n \leq \alpha_s} \left(\pi\left(\frac{p_{k_s} + p_{l_s}}{r_n}\right) - \pi\left(\frac{2s}{r_n}\right) \right) \leq$$

$$\leq F + E \cdot \sum_{n \in \mathbb{Z}^+, \, r_n \leq \alpha_s} \left(D_{\frac{1}{r_n}} \pi\left(p_{k_s} + p_{l_s}\right) - C_{\frac{1}{r_n}} \pi(2s) \right) \leq$$

$$\leq F + E \cdot \sum_{n \in \mathbb{Z}^+, \, r_n \leq \alpha_s} \left(D_{\frac{1}{r_n}} (k_s + l_s) - C_{\frac{1}{r_n}} \pi(2s) \right) =$$

$$= F + E \cdot \sum_{n \in \mathbb{Z}^+, \, r_n \leq \alpha_s} \left(D_{\frac{1}{r_n}} (2\,\pi(s) - a_s + b_s) - C_{\frac{1}{r_n}} \pi(2s) \right) =$$

$$= F + E \cdot \sum_{n \in \mathbb{Z}^+, \, r_n \leq \alpha_s} \left(D_{\frac{1}{r_n}} (\pi(2s) - (a_s - b_s - e_s)) - C_{\frac{1}{r_n}} \pi(2s) \right) =$$

$$= F + E \cdot \pi(2s) \cdot \sum_{n \in \mathbb{Z}^+, \, r_n \leq \alpha_s} \left(D_{\frac{1}{r_n}} - C_{\frac{1}{r_n}} \right),$$

where in the second inequality we have used that

$$\pi\left(p_{k_s} + p_{l_s}\right) \leq \pi\left(p_{k_s}\right) + \pi\left(p_{l_s}\right) = k_s + l_s.$$

ii) It is immediate from the previous item.

$\square$

When the series $\sum_{n \in \mathbb{Z}^+} \left(D_{1/r_n} - C_{1/r_n} \right)$ converges we have the optimal case. The following result is the alternate version of Theorem 7.0.22.

Theorem 7.0.23. *Let $\mathcal{P} = (p_i)_{i \in \mathbb{Z}^+}$ be an strictly increasing and u.d. sequence in $\mathbb{R}^+$.*

Let $a > 0$. Suppose that there exists $m_a \in \mathbb{Z}^+$ such that

1. There exists a constant $B_a > 0$ such that

$$p_{(a \cdot m)} \geq B_a \cdot p_m \text{ for every } m \in \mathbb{Z}, \ m \geq m_a.$$

2. For every $t > 0$ there exist constants $C_t, D_t > 0 \ \ C_t \leq D_t$, such that

$$D_t \cdot \pi(x) \geq \pi(t \cdot x) \geq C_t \cdot \pi(x) \text{ for every } x \geq p_{m_a}.$$

3. $\pi(x + y) \leq \pi(x) + \pi(y)$ for all $x, y \geq p_{m_a}$.

Let $s \in \mathbb{R}$, $s \geq p_{m_a}$, $s \notin \mathcal{P}$. We define $\alpha_s := \frac{3s - p_1}{p_1}$.

Let $a_s \in \{0, ..., \pi(s) - 1\}$ and $b_s \in \{1, ..., \pi(2s - p_1) - \pi(s)\}$ be such that

$$b_s \geq (a - 1)\pi(s) - a \cdot a_s.$$

Also suppose that the following three conditions are verified:
$a_s \leq \pi(s) - m_a$, $a_s - b_s - e_s = 0$ and $a_s < \pi(s) - \frac{\pi(2s)}{R}$, with $R := C_{(B_a + 1)} > 0$.

i) Assume that there exist constants $E > 0$, $F \geq 0$ and a sequence of positive numbers $(r_n)_{n \in \mathbb{Z}^+} \subseteq \mathbb{R}^+$ such that

$$\pi(x - y) \leq F + E \cdot \sum_{n=1}^{+\infty} (-1)^n \left(\pi\left(\frac{x}{r_n}\right) - \pi\left(\frac{y}{r_n}\right) \right) =$$

$$= F + E \cdot \sum_{n \in \mathbb{Z}^+, \ r_n \leq \frac{x}{p_1}} (-1)^n \left(\pi\left(\frac{x}{r_n}\right) - \tau\left(\frac{y}{r_n}\right) \right),$$

for all $x, y \in \mathbb{R}^+$, $2 p_1 \leq y \leq x$. Then

$$\pi\left((p_{k_s} + p_{l_s}) - 2s\right) \leq F + E \cdot \pi(2s) \cdot \sum_{n \in \mathbb{Z}^+, \ even \ n, \ r_n \leq \alpha_s} \left(D_{\frac{1}{r_n}} - C_{\frac{1}{r_n}} \right).$$

ii) Suppose that there exist constants $E > 0$, $F \geq 0$ and a sequence of positive numbers $(r_n)_{n \in \mathbb{Z}^+} \subseteq \mathbb{R}^+$ verifying that

$$\pi(x - y) \leq F + E \cdot \frac{\sum_{n=1}^{+\infty} (-1)^n \left(\pi \left(\frac{x}{r_n} \right) - \pi \left(\frac{y}{r_n} \right) \right)}{\pi(y)} =$$

$$= F + E \cdot \frac{\sum_{n \in \mathbb{Z}^+, \, r_n \leq \frac{x}{p_1}} (-1)^n \left(\pi \left(\frac{x}{r_n} \right) - \pi \left(\frac{y}{r_n} \right) \right)}{\pi(y)},$$

for all $x, y \in \mathbb{R}^+$, $2\, p_1 \leq y \leq x$. Then

$$\pi \left((p_{k_s} + p_{l_s}) - 2s \right) \leq F + E \cdot \sum_{n \in \mathbb{Z}^+, \, \text{even } n, \, r_n \leq \alpha_s} \left(D_{\frac{1}{r_n}} - C_{\frac{1}{r_n}} \right).$$

Proof. Observe that $2s < p_{k_s} + p_{l_s}$, by Lemma 5.0.15.

i)

$$\pi \left(p_{k_s} + p_{l_s} - 2s \right) \leq F + E \cdot \sum_{n \in \mathbb{Z}^+, \, r_n \leq \alpha_s} (-1)^n \left(\pi \left(\frac{p_{k_s} + p_{l_s}}{r_n} \right) - \pi \left(\frac{2s}{r_n} \right) \right) \leq$$

$$\leq F + E \cdot \sum_{n \in \mathbb{Z}^+, \, \text{even } n, \, r_n \leq \alpha_s} \left(\pi \left(\frac{p_{k_s} + p_{l_s}}{r_n} \right) - \pi \left(\frac{2s}{r_n} \right) \right) \leq$$

$$\leq F + E \cdot \sum_{n \in \mathbb{Z}^+, \, \text{even } n, \, r_n \leq \alpha_s} \left(D_{\frac{1}{r_n}} \, \pi \left(p_{k_s} + p_{l_s} \right) - C_{\frac{1}{r_n}} \, \pi \left(2s \right) \right) \leq$$

$$\leq F + E \cdot \sum_{n \in \mathbb{Z}^+, \, \text{even } n, \, r_n \leq \alpha_s} \left(D_{\frac{1}{r_n}} \left(k_s + l_s \right) - C_{\frac{1}{r_n}} \, \pi \left(2s \right) \right) =$$

$$= F + E \cdot \sum_{n \in \mathbb{Z}^+, \, \text{even } n, \, r_n \leq \alpha_s} \left(D_{\frac{1}{r_n}} \left(2\, \pi(s) - a_s + b_s \right) - C_{\frac{1}{r_n}} \, \pi \left(2s \right) \right) =$$

$$= F + E \cdot \sum_{n \in \mathbb{Z}^+, \, \text{even } n, \, r_n \leq \alpha_s} \left(D_{\frac{1}{r_n}} \left(\pi(2s) - (a_s - b_s - e_s) \right) - C_{\frac{1}{r_n}} \, \pi \left(2s \right) \right) =$$

$$= F + E \cdot \pi(2s) \cdot \sum_{n \in \mathbb{Z}^+, \, \text{even } n, \, r_n \leq \alpha_s} \left(D_{\frac{1}{r_n}} - C_{\frac{1}{r_n}} \right),$$

where in the third inequality we have used that

$$\pi \left(p_{k_s} + p_{l_s} \right) \leq \pi \left(p_{k_s} \right) + \pi \left(p_{l_s} \right) = k_s + l_s.$$

ii) It is immediate from the previous item.

$\square$

Definition 7.0.24. *Let $\mathcal{P} = (p_i)_{i \in \mathbb{Z}^+}$ and $(r_n)_{n \in \mathbb{Z}^+}$ be sequences in $\mathbb{R}^+$, with $\mathcal{P} = (p_i)_{i \in \mathbb{Z}^+}$ strictly increasing and uniformly discrete. We say that the distribution of $\mathcal{P} = (p_i)_{i \in \mathbb{Z}^+}$ is determined by $(r_n)_{n \in \mathbb{Z}^+}$ if the function*

$$\varphi_{\pi_{\mathcal{P}}} : \mathbb{R}^+ \to \mathbb{R}^{\mathbb{Z}^+}$$

defined by

$$\varphi_{\pi_{\mathcal{P}}}(x) := \left(\pi_{\mathcal{P}} \left(\frac{x}{r_n} \right) \right)_{n \in \mathbb{Z}^+} \quad \text{for all } x \in \mathbb{R}^+$$

is injective.

Next we will obtain new results on the approximation $p_{\pi(s)} + p_{\pi(s)+1} \approx 2s$ changing some hypotheses.

Chapter 8

On the approximation
$$p_{\pi(s)} + p_{\pi(s)+1} \approx 2s$$

In this chapter we will obtain new results over the approximation

$$p_{\pi(s)} + p_{\pi(s)+1} \approx 2s.$$

We will not consider the hypothesis

$$p_{(a \cdot m)} \geq B_a \cdot p_m \text{ for every } m \in \mathbb{Z}, \ m \geq m_a,$$

for given $a > 0$, but instead will have to suppose that the distribution function π is subadditive.

For sets with contractive distribution function we have the following result (see Definition 2.0.9).

Theorem 8.0.25. *Let $\mathcal{P} = (p_i)_{i \in \mathbb{Z}^+}$ be an strictly increasing and u.d. sequence in $\mathbb{R}^+$. Suppose that the distribution function π is subadditive. Let $s \in \mathbb{R}$ be such that $s \geq p_1$, $s \notin \mathcal{P}$.*

Let $a_s \in \{0, ..., \pi(s) - 1\}$ and $b_s \in \{1, ..., \pi(2s - p_1) - \pi(s)\}$. Suppose that there exist a constant $C > 0$ such that

$$\pi\left(|x - y|\right) \leq C \cdot |\pi(x) - \pi(y)|,$$

for all x, $y > 0$. Then

1. $\pi\left(|(p_{k_s} + p_{l_s}) - 2s|\right) \leq C \cdot (a_s + b_s).$

2. Hence $\pi\left(\left|\left(p_{\pi(s)} + p_{\pi(s)+1}\right) - 2s\right|\right) \leq C$, that is,

$$\left|\left(p_{\pi(s)} + p_{\pi(s)+1}\right) - 2s\right| < p_{([C]+1)}.$$

Proof. Observe that as $a_s \geq 0$ and $b_s \geq 1$, then $p_{k_s} \leq s$ and $p_{l_s} > s$.

1.

$$\pi\left(\left|\left(p_{k_s} + p_{l_s}\right) - 2s\right|\right) = \pi\left(\left|\left(p_{k_s} - s\right) + \left(p_{l_s} - s\right)\right|\right) \leq$$
$$\leq \pi\left(\left|p_{k_s} - s\right| + \left|p_{l_s} - s\right|\right) \leq \pi\left(\left|p_{k_s} - s\right|\right) + \pi\left(\left|p_{l_s} - s\right|\right) =$$
$$= \pi\left(s - p_{k_s}\right) + \pi\left(p_{l_s} - s\right) \leq C \cdot \left(\pi(s) - \pi\left(p_{k_s}\right)\right) + C \cdot \left(\pi\left(p_{l_s}\right) - \pi(s)\right) =$$
$$= C \cdot \left(l_s - k_s\right) = C \cdot \left(a_s + b_s\right).$$

where in the second inequality we have used that π is subadditive.

2. It is immediate from the previous item with $a_s = 0$ and $b_s = 1$.

$\square$

Theorem 8.0.26. *Let $\mathcal{P} = (p_i)_{i \in \mathbb{Z}^+}$ be an strictly increasing and u.d. sequence in $\mathbb{R}^+$. Suppose that the distribution function π is subadditive, and for every $t > 0$ there exist constants C_t, $D_t > 0$, $C_t \leq D_t$, verifying that*

$$D_t \cdot \pi(x) \geq \pi\left(t \cdot x\right) \geq C_t \cdot \pi(x) \text{ for every } x \geq p_1.$$

Let $s \in \mathbb{R}$, $s \geq p_1$, $s \notin \mathcal{P}$.
   Let $a_s \in \{0, ..., \pi(s) - 1\}$ and $b_s \in \{1, ..., \pi(2s - p_1) - \pi(s)\}$.
   Assume that there exist constants $E > 0$, $F \geq 0$, and a sequence of positive numbers $r := (r_n)_{n \in \mathbb{Z}^+} \subseteq \mathbb{R}^+$ such that

$$\pi\left(|x - y|\right) \leq F + E \cdot \sum_{n \in \mathbb{Z}^+, \, r_n \leq \max\left\{\left\lceil \frac{x}{p_1} \right\rceil, \left\lceil \frac{y}{p_1} \right\rceil\right\}} \frac{\left|\pi\left(\frac{x}{r_n}\right) - \pi\left(\frac{y}{r_n}\right)\right|}{\max\left\{\pi(x), \pi(y)\right\}}$$

for all $x, y > 0$. Then

1.

$$\pi\left(\left|\left(p_{k_s} + p_{l_s}\right) - 2s\right|\right) \leq 2F + E \cdot g_r(a_s, b_s, 2s),$$

with

$$g_r(a_s, b_s, 2s) := 2 \sum_{n \in \mathbb{Z}^+, \, r_n \leq \left\lceil \frac{s}{p_1} \right\rceil} \left(D_{\frac{1}{r_n}} - C_{\frac{1}{r_n}} \right) +$$

$$+ \sum_{n \in \mathbb{Z}^+, \, \left\lceil \frac{s}{p_1} \right\rceil < r_n \leq \left\lceil \frac{2s-p_1}{p_1} \right\rceil} \left(D_{\frac{1}{r_n}} - C_{\frac{1}{r_n}} \right) +$$

$$+ \frac{\sum_{n \in \mathbb{Z}^+, \, r_n \leq \left\lceil \frac{2s-p_1}{p_1} \right\rceil} D_{\frac{1}{r_n}} \cdot b_s + \sum_{n \in \mathbb{Z}^+, \, r_n \leq \left\lceil \frac{s}{p_1} \right\rceil} C_{\frac{1}{r_n}} \cdot a_s}{\pi(s)} .$$

2.

$$\pi \left(\left| \left(p_{\pi(s)} + p_{\pi(s)+1} \right) - 2s \right| \right) \leq 2F + E \cdot g_r(0, 1, 2s),$$

with

$$g_r(0, 1, 2s) = 2 \sum_{n \in \mathbb{Z}^+, \, r_n \leq \left\lceil \frac{s}{p_1} \right\rceil} \left(D_{\frac{1}{r_n}} - C_{\frac{1}{r_n}} \right) +$$

$$+ \sum_{n \in \mathbb{Z}^+, \, \left\lceil \frac{s}{p_1} \right\rceil < r_n \leq \left\lceil \frac{2s-p_1}{p_1} \right\rceil} \left(D_{\frac{1}{r_n}} - C_{\frac{1}{r_n}} \right) + \frac{\sum_{n \in \mathbb{Z}^+, \, r_n \leq \left\lceil \frac{2s-p_1}{p_1} \right\rceil} D_{\frac{1}{r_n}}}{\pi(s)} .$$

Proof. Observe that as $a_s \geq 0$ and $b_s \geq 1$, then $p_{k_s} \leq s$ and $p_{l_s} > s$. Besides $\pi(p_{l_s}) = l_s = \pi(s) + b_s > \pi(s)$, and $\pi(p_{k_s}) = k_s = \pi(s) - a_s \leq \pi(s)$. We also know that $p_{l_s} \leq 2s - p_1$.

i)

$$\pi\left(\left|(p_{k_s}+p_{l_s})-2s\right|\right)=\pi\left(\left|(p_{k_s}-s)+(p_{l_s}-s)\right|\right)\leq$$

$$\leq\pi\left(|p_{k_s}-s|+|p_{l_s}-s|\right)\leq\pi\left(|p_{k_s}-s|\right)+\pi\left(|p_{l_s}-s|\right)=$$

$$=\pi\left(s-p_{k_s}\right)+\pi\left(p_{l_s}-s\right)\leq$$

$$\leq F+E\cdot\sum_{n\in\mathbb{Z}^+,\,r_n\leq\max\left\{\left\lceil\frac{s}{p_1}\right\rceil,\,\left\lceil\frac{p_{k_s}}{p_1}\right\rceil\right\}}\frac{\left|\pi\left(\frac{s}{r_n}\right)-\pi\left(\frac{p_{k_s}}{r_n}\right)\right|}{\max\left\{\pi\left(p_{k_s}\right),\,\pi\left(s\right)\right\}}+$$

$$+F+E\cdot\sum_{n\in\mathbb{Z}^+,\,r_n\leq\max\left\{\left\lceil\frac{p_{l_s}}{p_1}\right\rceil,\,\left\lceil\frac{s}{p_1}\right\rceil\right\}}\frac{\left|\pi\left(\frac{p_{l_s}}{r_n}\right)-\pi\left(\frac{s}{r_n}\right)\right|}{\max\left\{\pi\left(p_{l_s}\right),\,\pi\left(s\right)\right\}}\leq$$

$$\leq 2F+\frac{E}{\pi(s)}\cdot\sum_{n\in\mathbb{Z}^+,\,r_n\leq\left\lceil\frac{s}{p_1}\right\rceil}\left(\pi\left(\frac{s}{r_n}\right)-\pi\left(\frac{p_{k_s}}{r_n}\right)\right)+$$

$$+\frac{E}{\pi(s)}\cdot\sum_{n\in\mathbb{Z}^+,\,r_n\leq\left\lceil\frac{p_{l_s}}{p_1}\right\rceil}\left(\pi\left(\frac{p_{l_s}}{r_n}\right)-\pi\left(\frac{s}{r_n}\right)\right)\leq$$

$$\leq 2F+\frac{E}{\pi(s)}\cdot\sum_{n\in\mathbb{Z}^+,\,r_n\leq\left\lceil\frac{s}{p_1}\right\rceil}\left(D_{\frac{1}{r_n}}\,\pi\left(s\right)-C_{\frac{1}{r_n}}\,\pi\left(p_{k_s}\right)\right)+$$

$$+\frac{E}{\pi(s)}\cdot\sum_{n\in\mathbb{Z}^+,\,r_n\leq\left\lceil\frac{2s-p_1}{p_1}\right\rceil}\left(D_{\frac{1}{r_n}}\,\pi\left(p_{l_s}\right)-C_{\frac{1}{r_n}}\,\pi\left(s\right)\right)=$$

$$=2F+\frac{E}{\pi(s)}\cdot\pi(s)\left(\left(\sum_{n\in\mathbb{Z}^+,\,r_n\leq\left\lceil\frac{s}{p_1}\right\rceil}D_{\frac{1}{r_n}}\right)-\left(\sum_{n\in\mathbb{Z}^+,\,r_n\leq\left\lceil\frac{2s-p_1}{p_1}\right\rceil}C_{\frac{1}{r_n}}\right)\right)+$$

$$+\frac{E}{\pi(s)}\cdot\sum_{n\in\mathbb{Z}^+,\,r_n\leq\left\lceil\frac{2s-p_1}{p_1}\right\rceil}D_{\frac{1}{r_n}}\cdot l_s-\frac{E}{\pi(s)}\cdot\sum_{n\in\mathbb{Z}^+,\,r_n\leq\left\lceil\frac{s}{p_1}\right\rceil}C_{\frac{1}{r_n}}\cdot k_s=$$

$$=2F+E\cdot\sum_{n\in\mathbb{Z}^+,\,r_n\leq\left\lceil\frac{s}{p_1}\right\rceil}D_{\frac{1}{r_n}}-E\cdot\sum_{n\in\mathbb{Z}^+,\,r_n\leq\left\lceil\frac{2s-p_1}{p_1}\right\rceil}C_{\frac{1}{r_n}}+$$

$$+\frac{E}{\pi(s)}\cdot\sum_{n\in\mathbb{Z}^+,\,r_n\leq\left\lceil\frac{2s-p_1}{p_1}\right\rceil}D_{\frac{1}{r_n}}\cdot b_s+E\cdot\sum_{n\in\mathbb{Z}^+,\,r_n\leq\left\lceil\frac{2s-p_1}{p_1}\right\rceil}D_{\frac{1}{r_n}}-$$

$$-E\cdot\sum_{n\in\mathbb{Z}^+,\,r_n\leq\left\lceil\frac{s}{p_1}\right\rceil}C_{\frac{1}{r_n}}+\frac{E}{\pi(s)}\sum_{n\in\mathbb{Z}^+,\,r_n\leq\left\lceil\frac{s}{p_1}\right\rceil}C_{\frac{1}{r_n}}\cdot a_s=$$

$$= 2F + E \cdot \left(\sum_{n\in\mathbb{Z}^+,\, r_n \leq \left\lceil \frac{s}{p_1} \right\rceil} \left(D_{\frac{1}{r_n}} - C_{\frac{1}{r_n}} \right) + \sum_{n\in\mathbb{Z}^+,\, r_n \leq \left\lceil \frac{2s-p_1}{p_1} \right\rceil} \left(D_{\frac{1}{r_n}} - C_{\frac{1}{r_n}} \right) \right) +$$

$$+ \frac{E}{\pi(s)} \sum_{n\in\mathbb{Z}^+,\, r_n \leq \left\lceil \frac{2s-p_1}{p_1} \right\rceil} D_{\frac{1}{r_n}} \cdot b_s + \frac{E}{\pi(s)} \sum_{n\in\mathbb{Z}^+,\, r_n \leq \left\lceil \frac{s}{p_1} \right\rceil} C_{\frac{1}{r_n}} \cdot a_s =$$

$$= 2F +$$

$$+ E \left(2 \sum_{n\in\mathbb{Z}^+,\, r_n \leq \left\lceil \frac{s}{p_1} \right\rceil} \left(D_{\frac{1}{r_n}} - C_{\frac{1}{r_n}} \right) + \sum_{n\in\mathbb{Z}^+,\, \left\lceil \frac{s}{p_1} \right\rceil < r_n \leq \left\lceil \frac{2s-p_1}{p_1} \right\rceil} \left(D_{\frac{1}{r_n}} - C_{\frac{1}{r_n}} \right) \right) +$$

$$+ E \cdot \frac{\sum_{n\in\mathbb{Z}^+,\, r_n \leq \left\lceil \frac{2s-p_1}{p_1} \right\rceil} D_{\frac{1}{r_n}} \cdot b_s + \sum_{n\in\mathbb{Z}^+,\, r_n \leq \left\lceil \frac{s}{p_1} \right\rceil} C_{\frac{1}{r_n}} \cdot a_s}{\pi(s)} =$$

$$= 2F + E \cdot g_r(a_s, b_s, 2s).$$

ii) It is immediate from the previous item taking $a_s = 0$ and $b_s = 1$.

$\square$

Chapter 9

Relationship between bounds type Hardy-Littlewood for the distribution function and the existence of an expression of a given number as sum of two elements of that set

[1]

9.1 Introduction

The development of the Additive Number Theory has enormously increased in recent years, motivated in a great part by the research on the distribution of the prime numbers, and especially on Goldbach's Conjecture. The contributions of B. Riemann, G. H. Hardy, J. E. Littlewood (see [9]), S. Ramanujan, I. M. Vinogradov (see [18], [19] and [20]), N. Chudakov, J. van der Corput (see [17]) and T. Estermann (see [6]) are classical and of the most importance. In addition, in 1930 Lev Schnirelmann proved the existence of a constant $C \in \mathbb{Z}^+$ such that every integer number greater than 1 can be expressed as

[1]2020 AMS Mathematics Subject Classification: 11A41, 11A25, 11P32, 11P82.

the sum of not more than C prime numbers. Schnirelmann also proved that $C < 8 \cdot 10^5$ (see [14] and [15]). In 1995 O. Ramaré reduced this bound to $C \leq 6$. In 2012 Terence Tao shew that $C \leq 5$ for the expressions of all the odd natural numbers greater than 1 (see [16]). In 2014 H. Helfgott proved the Weak Goldbach Conjecture (see [10]), obtaining in particular that $C \leq 4$. On the other hand, in 1973 Chen Jingrun proved that each sufficiently large even natual number can be expressed as the sum of either two prime numbers, or a prime number and a semiprime (this is, the product of two prime numbers; see [2]).

Other very important chapter in the Additive Number Theory is the problem of the expression of a number as sum of powers of natural numbers, and the particular case when these natural numbers are primes. In this part of the theory the contributions of Fermat, Euler, Lagrange and C. G. J. Jacobi are also classical.

We establish some notation. We denote $\mathbb{Z}^+ := \{m \in \mathbb{Z} \mid m > 0\}$. Given $A \subseteq \mathbb{R}^n$, we denote the indicator function of A with respect to $\mathbb{R}^n$ by χ_A, and define $xA := \{a\,x \mid a \in A\}$, $A + B := \{a + b \mid a \in A, \, b \in B\}$ for every $B \subseteq \mathbb{R}^n$. We denote by $\mathbb{P}$ the set of prime numbers, and $\mathbb{P}^* := \mathbb{P} \setminus \{2\}$. We denote the number of elements of a set A by $Card(A)$. We also denote the set of positive real numbers by $\mathbb{R}^+$. Given a real number x, we denote by $[x]$ its integer part.

9.1.1 Definitions

We work with the concepts of distribution function and discriminant function of a uniformly discrete set of real numbers $\mathcal{P}$. We also use other concepts which help us to determine whether a given real number is or not a sum of two elements of $\mathcal{P}$.

Definition 9.1.1 (Uniformly discrete sequence). *Let $\mathcal{P} = (p_i)_{i \in \mathbb{Z}^+}$ be a u.d. strictly increasing sequence in $\mathbb{R}$. We say that $\mathcal{P}$ is uniformly discrete (briefly, u.d.) if $\delta(\mathcal{P}) := \inf_{i \in \mathbb{Z}^+} \{p_{i+1} - p_i\} > 0$.*

Definition 9.1.2. *Let $\mathcal{P} = (p_i)_{i \in \mathbb{Z}^+}$ be a u.d. strictly increasing sequence in $\mathbb{R}$.*

1. *$a := p_1 = \min\{p_i \mid i \in \mathbb{Z}^+\} \in \mathbb{R}$.*

2. *$\mathbb{N}_{\geq a} := \{n \in \mathbb{N} \mid n \geq a\} \subseteq \mathbb{Z}^+$, $2\mathbb{N}_{\geq a} := \{2n \mid n \in \mathbb{N}, \, n \geq a\}$. We have analogous definitions for $\mathbb{R}_{\geq a}$ and $2\,\mathbb{R}_{\geq a}$.*

3. *The function $\pi := \pi_{\mathcal{P}} : \mathbb{R} \to \mathbb{R}$ defined by*

$$\pi(x) := Card\left(\{p \in \mathcal{P} \mid p \leq x\}\right) = Card\left(\mathcal{P} \cap (-\infty, \, x]\right),$$

for every $x \in \mathbb{R}$, is called the distribution function of $\mathcal{P}$. π is monotonically increasing.

Remark 9.1.3. *Observe these consequences of the definition of π:*

1. *For every $x \in \mathbb{R}^+$ we have $\left\{p_1, ..., p_{\pi(x)}\right\} \subseteq (-\infty, \, x]$, and then $p_{\pi(x)} \leq x$.*

2. *$\left(p \leq x \Leftrightarrow p \leq p_{\pi(x)}\right)$ for every $x \in \mathbb{R}^+$, $p \in \mathcal{P}$.*

Using Definition 9.1.2 we can formulate (Strong) Goldbach's Conjecture:

Conjecture 9.1.4 ((Strong) Goldbach's Conjecture).

$$2\mathbb{N}_{\geq 3} \subseteq \mathbb{P}^* + \mathbb{P}^*.$$

In 2014 Harald Andres Helfgott proved the Ternary (also called Odd) Goldbach's Conjecture, which we can formulate as follows:

Theorem 9.1.5 (Ternary Goldbach's Conjecture (see [10])).

$$2\mathbb{N}_{\geq 4} + 1 \subseteq \mathbb{P}^* + \mathbb{P}^* + \mathbb{P}^*.$$

This is, every odd integer greater or equal than 9 is the sum of three odd prime numbers.

Now we define the lower and upper Hardy-Littlewood conditions for a distribution function.

Definition 9.1.6 (Hardy-Littlewood Conditions). *Let $\mathcal{P} = (p_i)_{i \in \mathbb{Z}^+}$ be a u.d. strictly increasing sequence in $\mathbb{R}$. Let $\pi := \pi_{\mathcal{P}} : \mathbb{R} \to \mathbb{R}$ its distribution function.*

1. *We say that π verifies the upper Hardy-Littlewood condition if there exist constants $b_1, \, b_2 > 0$ and $b_3 \in \mathbb{R}$ such that*

$$\pi(x - y) \leq b_1 \, \pi(x) - b_2 \, \pi(y) + b_3 \text{ for all } x, \, y > 0, \, y \leq \frac{x}{2}.$$

2. *We say that π verifies the lower Hardy-Littlewood condition if there exist constants a_1, $a_2 > 0$ and $a_3 \in \mathbb{R}$ such that*

$$a_1\,\pi(x) - a_2\,\pi(y) + a_3 \leq \pi(x - y) \text{ for all } x,\, y > 0,\, y \leq \frac{x}{2}.$$

3. *π is said to verify the Hardy-Littlewood condition whether it verifies both the lower and the upper Hardy-Littlewood conditions.*

Next we define the Goldbach property of a uniformly discrete set of real numbers with respect to a given real number.

Definition 9.1.7 (Goldbach Property). *Let $\mathcal{P} = (p_i)_{i \in \mathbb{Z}^+}$ be a u.d. strictly increasing sequence in $\mathbb{R}$, and let $m \in \mathbb{R}$. We say that $\mathcal{P}$ verifies the Goldbach property for m if $2m \in \mathcal{P} + \mathcal{P}$.*

Let $\mathcal{P} = (p_i)_{i \in \mathbb{Z}^+}$ be a u.d. strictly increasing sequence in $\mathbb{R}$. Let $m \in \mathbb{R}_{\geq a}$, where $a = p_1$. We wonder if $2m \in \mathcal{P} + \mathcal{P}$. Obviously, if $m \in \mathcal{P}$ (the trivial case), then $2m = m + m \in \mathcal{P} + \mathcal{P}$ and the answer is affirmative. We want to know what happens if $m \notin \mathcal{P}$. Assume that $2m \in \mathcal{P} + \mathcal{P}$, with $m \notin \mathcal{P}$. There exist p, $q \in \mathcal{P}$, with $p \leq q$, such that $2m = p + q$. Then $2m \geq 2p$ and therefore $p \leq m$, or what is equivalent, $p \leq p_{\pi(m)}$. Thus we have that $p \in \mathcal{P}$, $2m - p \in \mathcal{P}$ and $p \leq p_{\pi(m)}$. Hence these last three conditions jointly are equivalent to $2m \in \mathcal{P} + \mathcal{P}$ when $m \notin \mathcal{P}$ (of course it is also true for $m \in \mathcal{P}$).

Definition 9.1.8 (Discriminant function (see [12])). *Let $\mathcal{P} = (p_i)_{i \in \mathbb{Z}^+}$ be a u.d. strictly increasing sequence in $\mathbb{R}$, and define $a := p_1$. The function $\psi : \mathbb{R}_{\geq a} \to \mathbb{N}$ defined by*

$$\psi(m) := \sum_{i=1}^{\pi(m)} \pi(2m - p_i) = \sum_{p \in \mathcal{P},\, p \leq m} \pi(2m - p) \text{ for all } m \in \mathbb{R}_{\geq a},$$

is called the discriminant function of $\mathcal{P}$.

The following result which justifies the name and the importance of the *discriminant fucntion.*

Lemma 9.1.9 (See [12]). *Let $\mathcal{P} = (p_i)_{i \in \mathbb{Z}^+}$ be a u.d. strictly increasing sequence in $\mathbb{R}$. Define $a := p_1$, and let $m \in \mathbb{R}_{\geq a}$, $m \notin \mathcal{P}$ be. Assume that the distribution function, π, of $\mathcal{P}$ verifies both that $\pi(m) = \pi(m - 1)$, and*

$$\pi(2m - p_i) - \pi(2(m - 1) - p_i) = \chi_{\mathcal{P}}(2m - p_i) \text{ for each } i \in \{1, ..., \pi(m - 1)\}.$$

Then:

1. $c(m) := \psi(m) - \psi(m-1) = \sum_{p \in \mathcal{P},\, p \le m} \chi_{\mathcal{P}}(2m - p) \ge 0$ *is the number of times that* $2m$ *can be expressed as sum of two elements of* $\mathcal{P}$ *(considering the same form* $p + q$ *and* $q + p$ *for all* $p, q \in \mathcal{P}$*).*

2. $c(m) = \psi(m) - \psi(m-1) > 0 \Leftrightarrow 2m \in \mathcal{P} + \mathcal{P}$.

Proof. Since $m \notin \mathcal{P}$, then $\pi(m) = \pi(m-1)$. Therefore:

$$\psi(m) - \psi(m-1) = \sum_{i=1}^{\pi(m)} \pi(2m - p_i) - \sum_{i=1}^{\pi(m-1)} \pi\left(2(m-1) - p_i\right) =$$

$$= \sum_{i=1}^{\pi(m-1)} \pi(2m - p_i) - \sum_{i=1}^{\pi(m-1)} \pi\left((2m - p_i) - 2\right) =$$

$$= \sum_{i=1}^{\pi(m-1)} \left[\pi(2m - p_i) - \pi\left((2m - p_i) - 2\right)\right] = \sum_{i=1}^{\pi(m-1)} \chi_{\mathcal{P}}(2m - p_i) =$$

$$= \sum_{i=1}^{\pi(m)} \chi_{\mathcal{P}}(2m - p_i) = \sum_{p \in \mathcal{P},\, p \le m} \chi_{\mathcal{P}}(2m - p) \ge 0,$$

where the fifth equality is consequence of the key assumption

$$\pi\left(2m - p_i\right) - \pi\left(2(m-1) - p_i\right) = \chi_{\mathcal{P}}\left(2m - p_i\right) \text{ for all } i \in \{1, ..., \pi(m-1)\}.$$

$\square$

Remark 9.1.10. *Observe that for the case* $\mathcal{P} = (p_i)_{i \in \mathbb{Z}^-} \subseteq \mathbb{Z}$ *the condition*

$$\pi\left(2m - p_i\right) - \pi\left(2(m-1) - p_i\right) = \chi_{\mathcal{P}}\left(2m - p_i\right) \text{ for each } i \in \{1, ..., \pi(m-1)\}$$

is equivalent to the one

$$2m - 1 - p_i \notin \mathcal{P} \text{ for every } i \in \{1, ..., \pi(m-1)\},$$

what is obviously verified when $\mathcal{P} \subseteq 2\mathbb{Z}$ *or* $\mathcal{P} \subseteq 2\mathbb{Z} + 1$, *and* $m \in \mathbb{Z}$.
    Besides notice that if $\mathcal{P} = (p_i)_{i \in \mathbb{Z}^+} \subseteq \mathbb{Z}$ *and* $m \in \mathbb{Z}$, *then*

$$m \notin \mathcal{P} \Leftrightarrow \pi(m) = \pi(m-1).$$

9.1.2   Results

Our main results are the following theorems.

Theorem 9.1.11. *Let* $\mathcal{P} = (p_i)_{i \in \mathbb{Z}^+} \subseteq \mathbb{R}$ *be a u.d. strictly increasing sequence and* $m \in \mathbb{R} \setminus \mathcal{P}$ *be such that* $m \geq p_1$, *and the distribution function,* π, *of* $\mathcal{P}$ *verifies both that* $\pi(m) = \pi(m-1)$, *and*

$$\pi(2m - p_i) - \pi(2(m-1) - p_i) = \chi_{\mathcal{P}}(2m - p_i) \text{ for each } i \in \{1, ..., \pi(m-1)\}.$$

Suppose that there exist constants a_1, a_2, b_1, $b_2 > 0$ *and* a_3, $b_3 \in \mathbb{R}$ *such that*

$$a_1 \pi(x) - a_2 \pi(y) + a_3 \leq \pi(x - y) \leq b_1 \pi(x) - b_2 \pi(y) + b_3 \text{ for all } x, y > 0, \ y \leq \frac{x}{2}.$$

Then

  1.

$$c(m) := \psi(m) - \psi(m-1) \geq (b_2 - a_2) \frac{\pi(m-1)(\pi(m-1) + 1)}{2} -$$
$$- (b_1 \pi(2m-2) - a_1 \pi(2m)) \pi(m-1) + (a_3 - b_3) \pi(m-1).$$

  2. Hence, if $\pi(2m) = \pi(2m-2)$, *then*

$$c(m) := \psi(m) - \psi(m-1) \geq (b_2 - a_2) \frac{\pi(m-1)(\pi(m-1) + 1)}{2} -$$
$$- (b_1 - a_1) \pi(m-1) \pi(2m-2) + (a_3 - b_3) \pi(m-1).$$

  3. Suppose that $\pi(m-1) \neq 0$ *and*

$$(b_2 - a_2)(\pi(m-1) + 1) > 2(b_1 \pi(2m-2) - a_1 \pi(2m)) - 2(a_3 - b_3).$$

  Then $c(m) > 0$, *what it is equivalent to* $2m \in \mathcal{P} + \mathcal{P}$.

  4. Assume that $\pi(m-1) \neq 0$, $\pi(2m) = \pi(2m-2)$ *and*

$$(b_2 - a_2)(\pi(m-1) + 1) > 2(b_1 - a_1)\pi(2m-2) - 2(a_3 - b_3).$$

  Then $2m \in \mathcal{P} + \mathcal{P}$.

Theorem 9.1.12. *Let $\mathcal{P} = (p_i)_{i \in \mathbb{Z}^+} \subseteq \mathbb{R}$ be a u.d. strictly increasing sequence and $m \in \mathbb{R} \setminus \mathcal{P}$ be such that $m \geq p_1$, and the distribution function, π, of $\mathcal{P}$ verifies both that $\pi(m) = \pi(m-1)$, and*

$$\pi(2m - p_i) - \pi(2(m-1) - p_i) = \chi_{\mathcal{P}}(2m - p_i) \text{ for each } i \in \{1, ..., \pi(m-1)\}.$$

Suppose that there exist constants a_1, a_2, b_1, $b_2 > 0$ and a_3, $b_3 \in \mathbb{R}$ such that

$$a_1 \pi(x) - a_2 \pi(y) + a_3 \leq \pi(x-y) \leq b_1 \pi(x) - b_2 \pi(y) + b_3 \text{ for all } x, y > 0, \; y \leq \frac{x}{2}.$$

Then

1.

$$c(m) = \psi(m) - \psi(m-1) \leq (a_2 - b_2) \frac{\pi(m-1)\,(\pi(m-1)+1)}{2} -$$
$$- (a_1\,\pi(2m-2) - b_1\,\pi(2m))\,\pi(m-1) + (b_3 - a_3)\,\pi(m-1).$$

2. *Hence, if $\pi(2m) = \pi(2m-2)$, then*

$$c(m) = \psi(m) - \psi(m-1) \leq (a_2 - b_2) \frac{\pi(m-1)\,(\pi(m-1)+1)}{2} -$$
$$- (a_1 - b_1)\,\pi(m-1)\,\pi(2m-2) + (b_3 - a_3)\,\pi(m-1).$$

3. *Assume that $\pi(m-1) \neq 0$ and*

$$(b_2 - a_2)\,(\pi(m-1)+1) = 2\,(b_1\,\pi(2m) - a_1\,\pi(2m-2)) + 2\,(b_3 - a_3).$$

   Then $c(m) = 0$, what it is equivalent to $2m \notin \mathcal{P} + \mathcal{P}$.

4. *Suppose that $\pi(m-1) \neq 0$, $\pi(2m) = \pi(2m-2)$ and*

$$(b_2 - a_2)\,(\pi(m-1)+1) = 2\,(b_1 - a_1)\,\pi(2m-2) + 2\,(b_3 - a_3).$$

   Then $2m \notin \mathcal{P} + \mathcal{P}$.

Remark 9.1.13. *Let $\mathcal{P} = (p_i)_{i \in \mathbb{Z}^+} \subseteq \mathbb{R}$ be a u.d. strictly increasing sequence, and let π be the distribution function of $\mathcal{P}$. Let b_1, $b_2 > 0$ and $b_3 \in \mathbb{R}$ be. The following statements are equivalent:*

1. $\pi(x - y) \leq b_1\,\pi(x) - b_2\,\pi(y) + b_3$ for all x, $y \in \mathbb{R}$, $y \leq \frac{x}{2}$.

2. $b_1\,\pi(z + y) \geq \pi(z) + b_2\,\pi(y) - b_3$ for every y, $z \in \mathbb{R}$, $y \leq z$.

3. $\pi(z + y) \geq \frac{1}{b_1}\,\pi(z) + \frac{b_2}{b_1}\,\pi(y) - \frac{b_3}{b_1}$ for every y, $z \in \mathbb{R}$, $y \leq z$.

Remark 9.1.14. *Let $\mathcal{P} = (p_i)_{i \in \mathbb{Z}^+} \subseteq \mathbb{R}$ be a u.d. strictly increasing sequence, and let π be the distribution function of $\mathcal{P}$. Let a_1, $a_2 > 0$ and $a_3 \in \mathbb{R}$ be. The next statements are equivalent:*

1. $a_1\,\pi(x) - a_2\,\pi(y) + a_3 \leq \pi(x - y)$ for each x, $y \in \mathbb{R}$, $y \leq \frac{x}{2}$.

2. $a_1\,\pi(z + y) \leq \pi(z) + a_2\,\pi(y) - a_3$ for every y, $z \in \mathbb{R}$, $y \leq z$.

3. $\pi(z + y) \leq \frac{1}{a_1}\,\pi(z) + \frac{a_2}{a_1}\,\pi(y) - \frac{a_3}{a_1}$ for all y, $z \in \mathbb{R}$, $y \leq z$.

The paper is structured as follows. Section 1 contains definitions and the main results. In section 2 we recall some well known results connected with the Second Hardy-Littlewood Conjecture in order to motivate our results. In section 3 we prove Theorem 9.1.11 and we obtain several consequences which may be applied to Goldbach's Conjecture for uniformly discrete sequences. Section 4 is devoted to the proof of Theorem 9.1.12, and we also obtain some consequences. In section 5 we generalize the Hardy-Littlewood conditions and obtain analogous results to Theorem 9.1.11 and Theorem 9.1.12. Finally in section 6 we obtain sufficient conditions in order to know whether a given even natural number is a sum of two prime numbers using induction on the number of prime factors of the given number.

9.2   Motivation: the prime number set.

Consider the prime number set $\mathbb{P}$, and its distribution function π. The Second Hardy-Littlewood Conjecture precisely states that the distribution function of the prime numbers set is subadditive (see [9] and [7]). That is:

$$\pi(x + y) \leq \pi(x) + \pi(y) \text{ for all } x, y \in \mathbb{R}, y \leq x.$$

We obtain the following results from [4]:

Remark 9.2.1.

i) $\pi(x+y) \le \pi(x)+\pi(y)$ *for all* x, $y \in \mathbb{R}$, $1 \le x \le y \le \frac{7}{5} x \log x \log_2 x$. *(See Theorem 2.6, p. 73).*

ii) $\pi(x+y) \le \pi(x)+\pi(y)$ *for all* x, $y \in [3, +\infty)$ *such that* $\frac{1}{109} \le \frac{y}{x} \le 1$. *(See Theorem 2.5, p. 67).*

iii) $\frac{x}{\log x - 1} \le \pi(x) \le \frac{x}{\log x - 1.1}$ *for each* $x \in [60184, +\infty)$, *where the first inequality is true even for every* $x \in [5393, +\infty)$. *(See Theorem 1.10, p. 37, and Theorem 6.9 of [5]). From these inequalities we deduce that*

$$\pi(a\,x) \ge a\,\frac{1 - \frac{1.1}{\log 60184}}{1 + \alpha}\,\pi(x)$$

for all $\alpha \in \mathbb{R}^+$, $a \in [1, 60184^\alpha]$, $x \in [60184, +\infty)$; *and*

$$\pi(a\,x) \le a\,\frac{\log(60184) - 1}{\log(60184) - 1.1}\,\pi(x) \le 1.010095746224 \cdot a \cdot \pi(x)$$

for every $a \in [1, +\infty)$, $x \in [60184, +\infty)$.

Rosser and Schoendfeld proved that $\pi(2x) \le 2\,\pi(x)$ for all $x \in [3, +\infty)$. Moreover, they proved that $\pi(2x) < 2\,\pi(x)$ for $x \ge 11$ (see [13], and also [11]).

Examples 9.2.2.

i) $\pi(2x) \ge 1.693424\,\pi(x)$ *for all* $x \in [60184, +\infty)$ *(taking* $\alpha = 0.06299$*)*.

ii) $\pi(3x) \ge 2.45505\,\pi(x)$ *for all* $x \in [60184, +\infty)$ *(taking* $\alpha = 0.09983$*)*.

Remark 9.2.3. *From [5] we have the inequalities*

$$\frac{x}{\log x}\left(1 + \frac{1}{\log x} + \frac{2}{\log^2 x}\right) \le \pi(x) \le \frac{x}{\log x}\left(1 + \frac{1}{\log x} + \frac{2.334}{\log^2 x}\right),$$

where the first inequality is verified for all $x \ge 88783$ *and the second one is for all* $x \ge 2953652287$.

Remark 9.2.4. *Let* x, $y \ge 0$, $x \le y$. *As the distribution function* π *is monotonically increasing, then*

$$\pi(x) \le \pi(x+y),\ \pi(y) \le \pi(x+y).$$

Thus $\pi(x) + \pi(y) \le 2\,\pi(x+y)$, *and then*

$$\frac{1}{2}\,\pi(x) + \frac{1}{2}\,\pi(y) \le \pi(x+y).$$

Obviously this gross lower bound is true for the distribution function of whatever uniformly discrete set of real numbers.

9.3   Proof of Theorem 9.1.11 and consequences

In this section we will prove Theorem 9.1.11 and obtain some consequences.

Proof of Theorem 9.1.11. We will prove the first item. The other items are immediate consequences of the first one.

$$c(m) = \psi(m) - \psi(m-1) = \sum_{i=1}^{\pi(m-1)} \pi(2m - p_i) - \sum_{i=1}^{\pi(m-1)} \pi(2(m-1) - p_i) \geq$$

$$\geq a_1 \sum_{i=1}^{\pi(m-1)} \pi(2m) - a_2 \sum_{i=1}^{\pi(m-1)} \pi(p_i) - b_1 \sum_{i=1}^{\pi(m-1)} \pi(2(m-1)) +$$

$$+ b_2 \sum_{i=1}^{\pi(m-1)} \pi(p_i) + a_3 \pi(m-1) - b_3 \pi(m-1) = a_1 \pi(m-1) \pi(2m) -$$

$$- a_2 \frac{\pi(m-1)(\pi(m-1)+1)}{2} - b_1 \pi(m-1) \pi(2m-2) +$$

$$+ b_2 \frac{\pi(m-1)(\pi(m-1)+1)}{2} + (a_3 - b_3) \pi(m-1) =$$

$$= (b_2 - a_2) \frac{\pi(m-1)(\pi(m-1)+1)}{2} + a_1 \pi(m-1) \pi(2m) -$$

$$- b_1 \pi(m-1) \pi(2m-2) + (a_3 - b_3) \pi(m-1) =$$

$$= (b_2 - a_2) \frac{\pi(m-1)(\pi(m-1)+1)}{2} - (b_1 \pi(2m-2) - a_1 \pi(2m)) \pi(m-1) +$$

$$+ (a_3 - b_3) \pi(m-1),$$

where we have used that $\pi(p_i) = i$ for every $i \in \mathbb{Z}^+$. $\qquad\square$

The next observation is crucial because it tells us that the Hardy-Littlewood bounds for the distribution function do not allow us to identify the sequence $\mathcal{P}$.

Remark 9.3.1. *Let $m \in \mathbb{Z}^+$, $n \in \{0, ..., m-1\}$ be. Consider*

$$\mathcal{P} := \mathcal{P}_{m,n} := m\,\mathbb{N} + n = \{mt + n \; : \; t \in \mathbb{N}\},$$

and its distribution function $\pi : \mathbb{R}^+ \to \mathbb{R}$, which is given by

$$\pi(x) := \left\lceil \frac{x+n}{m} \right\rceil \quad \text{for every } x \in \mathbb{R}^+.$$

A direct calculus shows that for all x, $y \in \mathbb{R}^+$ *we have that*

$$\pi\left(x+y\right) < \pi\left(x\right) + \pi\left(y\right) + 2$$
$$\pi\left(x+y\right) > \pi\left(x\right) + \pi\left(y\right) - 1.$$

Obviously this is equivalent to

$$\pi\left(x+y\right) \leq \pi\left(x\right) + \pi\left(y\right) + 1$$
$$\pi\left(x+y\right) \geq \pi\left(x\right) + \pi\left(y\right),$$

for every x, $y \in \mathbb{R}^+$.

Observe that the (lower and upper) bounds type Hardy-Littlewood that we have obtained are independent of the values of m and n. In other words, these bounds are the same for all the arithmetic progressions. This "universality" is the reason why Theorem 9.1.11 and Theorem 9.1.12 do not work well for arithmetic progressions.

Indeed, by Remark 9.1.13 and Remark 9.1.14, we obtain that

$$a_1 = b_1 = a_2 = b_2 = 1,\ a_3 = -1,\ b_3 = 0.$$

These values do not give us information neither from Theorem 9.1.11 nor Theorem 9.1.12.

From Theorem 9.1.11 we obtain the following consequences.

Corollary 9.3.2. *Let* $\mathcal{P} = (p_i)_{i \in \mathbb{Z}^+} \subseteq \mathbb{R}$ *be a u.d. strictly increasing sequence and* $m \in \mathbb{R} \setminus \mathcal{P}$ *be such that* $m \geq p_1$, *and the distribution function,* π, *of* $\mathcal{P}$ *verifies both that* $\pi\left(m\right) = \pi\left(m-1\right)$, *and*

$$\pi\left(2m - p_i\right) - \pi\left(2\left(m-1\right) - p_i\right) = \chi_{\mathcal{P}}\left(2m - p_i\right)\ \textit{for each } i \in \left\{1, ..., \pi\left(m-1\right)\right\}.$$

Suppose that there exist constants a_1, a_2, b_1, $b_2 > 0$ *and* a_3, $b_3 \in \mathbb{R}$ *such that*

$$a_1\pi(x) - a_2\pi(y) + a_3 \leq \pi(x-y) \leq b_1\pi(x) - b_2\pi(y) + b_3\ \textit{for all } x, y \in \mathbb{R},\ y \leq \frac{x}{2}.$$

Suppose that $2m \notin \mathcal{P} + \mathcal{P}$, *that is,* $c(m) = 0$. *Then*

1. *If* $\pi\left(m-1\right) \neq 0$, *then*

$$\left(b_1\ \pi\left(2m - 2\right) - a_1\ \pi\left(2m\right)\right) - \left(a_3 - b_3\right) \geq \left(b_2 - a_2\right)\ \frac{\pi\left(m-1\right) + 1}{2}.$$

2. In particular, if $\pi(2m) = \pi(2m - 2)$ and $\pi(m - 1) \neq 0$, then

$$(b_1 - a_1)\,\pi(2m - 2) - (a_3 - b_3) \geq (b_2 - a_2)\,\frac{\pi(m - 1) + 1}{2}.$$

Corollary 9.3.3. *Let $\mathcal{P} = (p_i)_{i \in \mathbb{Z}^+} \subseteq \mathbb{R}$ be a u.d. strictly increasing sequence with distribution function π. Suppose that there exist constants $a_1, a_2, b_1, b_2 > 0$ and $a_3, b_3 \in \mathbb{R}$ such that*

$$a_1\pi(x) - a_2\pi(y) + a_3 \leq \pi(x - y) \leq b_1\pi(x) - b_2\pi(y) + b_3 \text{ for all } x, y \in \mathbb{R},\ y \leq \frac{x}{2}.$$

Suppose that there exists a sequence S divergent to $+\infty$ of elements $m \in \mathbb{R}^+ \backslash \mathcal{P}$, $m \geq p_1$, with $2m \notin \mathcal{P} + \mathcal{P}$ and verifying both that $\pi(m) = \pi(m - 1) \neq 0$, and

$$\pi(2m - p_i) - \pi(2(m - 1) - p_i) = \chi_{\mathcal{P}}(2m - p_i) \text{ for each } i \in \{1, ..., \pi(m - 1)\}.$$

Also assume that

- $\lim_{x \to +\infty} \pi(x) = +\infty.$

- $L := \liminf_{x \to +\infty} \frac{\pi(x - 1)}{\pi(2x - 2)} \in (0, +\infty).$

- *There exists $C > 0$ such that $\pi(2m) - \pi(2m - 2) \leq C.$*

Then

$$b_1 - a_1 \geq \frac{L}{2}\,(b_2 - a_2).$$

Proof. It is an immediate result from the third item of Corollary 9.3.2, dividing all the expression by $\pi(2m - 2) > 0$, and then taking lower limits when $m \to +\infty$. $\qquad\square$

From Corollary 9.3.3 we obtain the following consequence inspired in the sequence of the prime numbers (see the first proofs of Number Prime Theorem in [8] and [3]).

Corollary 9.3.4. *Let $\mathcal{P} = (p_i)_{i \in \mathbb{Z}^+} \subseteq \mathbb{R}$ be a u.d. strictly increasing sequence and let π its distribution function. Assume that there exist constants $b_1, b_2 > 0$ and $a_3, b_3 \in \mathbb{R}$ such that*

$$\pi(x) - \pi(y) + a_3 \leq \pi(x - y) \leq b_1\,\pi(x) - b_2\,\pi(y) + b_3 \text{ for all } x, y \in \mathbb{R},\ y \leq \frac{x}{2}.$$

Suppose that there exists a sequence S divergent to $+\infty$ of elements $m \in \mathbb{R}^+ \setminus \mathcal{P}$ with $m \geq p_1$, $2m \notin \mathcal{P} + \mathcal{P}$, and verifying that verifies both that $\pi(m) = \pi(m-1) \neq 0$, and

$$\pi(2m - p_i) - \pi(2(m-1) - p_i) = \chi_{\mathcal{P}}(2m - p_i) \ \text{for each } i \in \{1, ..., \pi(m-1)\}.$$

Also assume that

- $\lim_{x \to +\infty} \pi(x) = +\infty$.

- $L := \liminf_{x \to +\infty} \frac{\pi(x-1)}{\pi(2x-2)} = \frac{1}{2}$.

- *There exists $C > 0$ such that $\pi(2m) - \pi(2m-2) \leq C$.*

Then

$$4\, b_1 - b_2 \geq 3.$$

Proof. This result follows directly from Corollary 9.3.3 taking $a_1 = 1 = a_2$ and $L = \frac{1}{2}$. $\qquad\square$

Remark 9.3.5. *Corollary 9.3.4 tells us that under its hypotheses, if $4\,b_1 - b_2 < 3$, then the Goldbach's Conjecture is true for the set $\mathcal{P}$ since some number m_0.*

Lemma 9.3.6. *Let $\mathcal{P} = (p_i)_{i \in \mathbb{Z}^+} \subseteq \mathbb{R}$ be a u.d. strictly increasing sequence, and let π be its distribution function. Let b_1, $b_2 > 0$ and $b_3 \in \mathbb{R}$ be. Suppose that*

1. $\pi(x - y) \leq b_1\, \pi(x) - b_2\, \pi(y) + b_3$ for every $x, y \in \mathbb{R}$, $y \leq \frac{x}{2}$.

2. $\lim_{x \to +\infty} \pi(x) = +\infty$.

3. $M := \limsup_{x \to +\infty} \frac{\pi(x)}{\pi\left(\frac{x}{2}\right)} \in (0, +\infty)$.

Then $1 + b_2 \leq M\, b_1$.

Proof. Let $x > 0$. Define $y := \frac{x}{2}$. Then

$$\pi\left(\frac{x}{2}\right) = \pi\left(x - \frac{x}{2}\right) \leq b_1\, \pi(x) - b_2\, \pi\left(\frac{x}{2}\right) + b_3.$$

Hence

$$(1 + b_2)\, \pi\left(\frac{x}{2}\right) \leq b_1\, \pi(x) + b_3.$$

Then

$$1 + b_2 \leq b_1 \frac{\pi(x)}{\pi\left(\frac{x}{2}\right)} + \frac{b_3}{\pi\left(\frac{x}{2}\right)} \text{ for each } x > 0.$$

Taking limits when $x \to +\infty$ we obtain $1 + b_2 \leq M\, b_1$. $\square$

With an analogous proof we have the following result for our usual functional lower bound of $\pi(x-y)$.

Lemma 9.3.7. *Let $\mathcal{P} = (p_i)_{i \in \mathbb{Z}^+} \subseteq \mathbb{R}$ be a u.d. strictly increasing sequence, and let π be its distribution function. Let $a_1,\ a_2 > 0$ and $a_3 \in \mathbb{R}$ be. Assume that*

1. $a_1\, \pi(x) - a_2\, \pi(y) + a_3 \leq \pi(x-y)$ *for each* $x,\ y \in \mathbb{R},\ y \leq \frac{x}{2}$.

2. $\lim_{x \to +\infty} \pi(x) = +\infty$.

3. $M' := \liminf_{x \to +\infty} \frac{\pi(x)}{\pi\left(\frac{x}{2}\right)} \in (0,\ +\infty)$.

Then $1 + a_2 \geq M'\, a_1$.

9.4   Proof of Theorem 9.1.12 and consequences

In this section we will prove Theorem 9.1.12.

Proof of Theorem 9.1.12. We will prove the first item. The other items follows directly from the first one.

$$c(m) = \psi(m) - \psi(m-1) = \sum_{i=1}^{\pi(m-1)} \pi(2m - p_i) - \sum_{i=1}^{\pi(m-1)} \pi(2(m-1) - p_i) \leq$$

$$\leq b_1 \sum_{i=1}^{\pi(m-1)} \pi(2m) - b_2 \sum_{i=1}^{\pi(m-1)} \pi(p_i) - a_1 \sum_{i=1}^{\pi(m-1)} \pi(2(m-1) - p_i) +$$

$$+ a_2 \sum_{i=1}^{\pi(m-1)} \pi(p_i) + b_3 \pi(m-1) - a_3 \pi(m-1) = b_1 \pi(m-1) \pi(2m) -$$

$$- b_2 \frac{\pi(m-1)(\pi(m-1)+1)}{2} - a_1 \pi(m-1) \pi(2m-2) +$$

$$+ a_2 \frac{\pi(m-1)(\pi(m-1)+1)}{2} + (b_3 - a_3) \pi(m-1) =$$

$$= (a_2 - b_2) \frac{\pi(m-1)(\pi(m-1)+1)}{2} + b_1 \pi(m-1) \pi(2m) -$$

$$- a_1 \pi(m-1) \pi(2m-2) + (b_3 - a_3) \pi(m-1) =$$

$$= (a_2 - b_2) \frac{\pi(m-1)(\pi(m-1)+1)}{2} - (a_1 \pi(2m-2) - b_1 \pi(2m)) \pi(m-1) +$$

$$+ (b_3 - a_3) \pi(m-1).$$

$\square$

From Theorem 9.1.12 we obtain the following consequences.

Corollary 9.4.1. *Let $\mathcal{P} = (p_i)_{i \in \mathbb{Z}^+} \subseteq \mathbb{R}$ be a u.d. strictly increasing sequence and $m \in \mathbb{R} \setminus \mathcal{P}$ be such that $m \geq p_1$, and the distribution function, π, of $\mathcal{P}$ verifies both that $\pi(m) = \pi(m-1) \neq 0$, and*

$$\pi(2m - p_i) - \pi(2(m-1) - p_i) = \chi_{\mathcal{P}}(2m - p_i) \text{ for each } i \in \{1, ..., \pi(m-1)\}.$$

Suppose that there exist constants a_1, $a_2 > 0$ and $b_3 \in \mathbb{R}$ such that

$$\pi(x - y) = a_1 \pi(x) - a_2 \pi(y) + a_3 \text{ for all } x, y > 0, y \leq \frac{x}{2}.$$

Then $2m \notin \mathcal{P} + \mathcal{P}$.

Proof. It is an immediate consequence of the fourth item of Theorem 9.1.12.

$\square$

Remark 9.4.2. *Observe that the following conditions are equivalent:*

- *There exist constants a_1, $a_2 > 0$ and $b_3 \in \mathbb{R}$ such that*

$$\pi(x - y) = a_1\, \pi(x) - a_2\, \pi(y) + a_3 \text{ for all } x,\ y > 0,\ y \leq \frac{x}{2}.$$

- *There exist constants c_1, $c_2 > 0$ and $c_3 \in \mathbb{R}$ such that*

$$\pi(x + y) = c_1\, \pi(x) + c_2\, \pi(y) + c_3 \text{ for all } x,\ y > 0,\ y \leq x.$$

Corollary 9.4.3. *Let $\mathcal{P} = (p_i)_{i \in \mathbb{Z}^+} \subseteq \mathbb{R}$ be a u.d. strictly increasing sequence with distribution function π. Assume that there exist constants $a_1, a_2, b_1, b_2 > 0$ and $a_3,\ b_3 \in \mathbb{R}$ such that*

$$a_1\, \pi(x) - a_2\, \pi(y) + a_3 \leq \pi(x - y) \leq b_1\, \pi(x) - b_2\, \pi(y) + b_3 \text{ for all } x, y \in \mathbb{R},\ y \leq \frac{x}{2}.$$

Assume that there exists a sequence S divergent to $+\infty$ of elements $m \in \mathbb{R}^+ \setminus \mathcal{P}$ such that $m \geq p_1$, $2m \in \mathcal{P} + \mathcal{P}$, $\pi(m) = \pi(m - 1) \neq 0$ and

$$\pi(2m - p_i) - \pi(2(m - 1) - p_i) = \chi_{\mathcal{P}}(2m - p_i) \text{ for each } i \in \{1, ..., \pi(m - 1)\}.$$

Then

$$b_1\, \pi(2m) - a_1\, \pi(2m - 2) + b_3 - a_3 > \frac{b_2 - a_2}{2}\, (\pi(m - 1) + 1),$$

for every $m \in S$. Assume that the following conditions are also satisfied:

i) $\lim_{x \to +\infty} \pi(x) = +\infty$.

ii) $L := \liminf_{x \to +\infty} \frac{\pi(x-1)}{\pi(2x-2)} \in (0,\ +\infty)$.

iii) There exists $C > 0$ such that $\pi(2m) - \pi(2m - 2) \leq C$.

Then

$$b_1 - a_1 \geq \frac{L}{2}\, (b_2 - a_2).$$

Proof. We obtain

$$b_1 \frac{\pi(2m)}{\pi(2m-2)} - a_1 + \frac{b_3 - a_3}{\pi(2m-2)} > \frac{b_2 - a_2}{2} \left(\frac{\pi(m-1)}{\pi(2m-2)} + \frac{1}{\pi(2m-2)} \right),$$

for every $m \in S$, $m \geq m_0$.

Taking now lower limits when $m \to +\infty$, $m \in S$, we finally have that

$$b_1 - a_1 \geq \frac{L}{2}(b_2 - a_2).$$

$\square$

Remark 9.4.4. *If in Corollary 9.4.3 we take in particular $a_1 = 1 = a_2$, $L = \frac{1}{2}$, then*

$$4\,b_1 - b_2 \geq 3.$$

Corollary 9.4.5. *Let $\mathcal{P} = (p_i)_{i \in \mathbb{Z}^+} \subseteq \mathbb{R}$ be a u.d. strictly increasing sequence and $m \in \mathbb{R} \setminus \mathcal{P}$ be such that $m \geq p_1$, and the distribution function, π, of $\mathcal{P}$ verifies both that $\pi(m) = \pi(m-1)$, and*

$$\pi(2m - p_i) - \pi(2(m-1) - p_i) = \chi_\mathcal{P}(2m - p_i) \text{ for each } i \in \{1, ..., \pi(m-1)\}.$$

Suppose that there exist constants a_1, a_2, b_1, $b_2 > 0$, with $a_1 \neq b_1$, verifying that

$$a_1\,\pi(x) - a_2\,\pi(y) \leq \pi(x-y) \leq b_1\,\pi(x) - b_2\,\pi(y) \text{ for all } x,\, y > 0,\, y \leq \frac{x}{2}.$$

Assume that

1. $\pi(m-1) \neq 0$.

2. $\pi(2m) = \pi(2m-2)$.

3.
$$\frac{b_2 - a_2}{2(b_1 - a_1)} = \frac{\pi(2m-2)}{\pi(m-1) + 1}.$$

Then $2m \notin \mathcal{P} + \mathcal{P}$.

Proof. This result follows immediately from the fourth item of Theorem 9.1.12.

$\square$

Remark 9.4.6. *Observe that Corollary 9.4.5 tells us nothing on u.d. strictly increasing sequence $\mathcal{P} = (p_i)_{i \in \mathbb{Z}^+} \subseteq \mathbb{R}$ verifying both that*

1. *There exists $x_0 \in \mathbb{R}$ such that*

$$\pi(2x) < 2\left(\pi(x) + 1\right) \text{ for every } x \in \mathbb{R}, \ x \geq x_0.$$

2. *$\liminf_{x \to +\infty} \frac{\pi(2x-2)}{\pi(x-1)+1} = 2.$*

As we know, this is the case of the prime number set, $\mathbb{P}$ (see [13] for the first item, and [3] and [8] for the second one; also see Section 2).

9.5   Generalized Hardy-Littlewood Conditions

In this section we will generalize the Hardy-Littlewood conditions and obtain analogous results to both Theorem 9.1.11 and Theorem 9.1.12.

Theorem 9.5.1. *Let $\mathcal{P} = (p_i)_{i \in \mathbb{Z}^+} \subseteq \mathbb{R}$ be a u.d. strictly increasing sequence and $m \in \mathbb{R} \setminus \mathcal{P}$ be such that $m \geq p_1$, and the distribution function, π, of $\mathcal{P}$ verifies both that $\pi(m) = \pi(m-1)$, and*

$$\pi(2m - p_i) - \pi(2(m-1) - p_i) = \chi_{\mathcal{P}}(2m - p_i) \text{ for each } i \in \{1, ..., \pi(m-1)\}.$$

Suppose that there exist constants $a_1, a_2, a_5, b_1, b_2, b_5 > 0$ and $a_3, a_4, b_3, b_4 \in \mathbb{R}$ such that

$$a_1 \pi(x) - a_2 \pi(y) + a_3 + a_4 x - a_5 y \leq \pi(x-y) \leq b_1 \pi(x) - b_2 \pi(y) + b_3 + b_4 x - b_5 y$$

for all $x, y > 0$, $y \leq \frac{x}{2}$.
    Then

1.

$$c(m) = \psi(m) - \psi(m-1) \geq (b_2 - a_2) \frac{\pi(m-1)\left(\pi(m-1)+1\right)}{2} -$$

$$- \left(b_1 \pi(2m-2) - a_1 \pi(2m)\right) \pi(m-1) + (a_3 - b_3) \pi(m-1) -$$

$$- \left(b_4(2m-2) - 2a_4 m\right) \pi(m-1) + (b_5 - a_5) \sum_{i=1}^{\pi(m-1)} p_i.$$

2. Suppose that

$$(b_2 - a_2) \, \frac{\pi \, (m-1) + 1}{2} + (b_5 - a_5) \, \frac{\sum_{i=1}^{\pi(m-1)} p_i}{\pi \, (m-1)} >$$

$$> b_1 \, \pi \, (2m-2) - a_1 \, \pi \, (2m) + 2m \, (b_4 - a_4) + b_3 - a_3 - 2b_4.$$

Then $c(m) > 0$, that is, $2m \in \mathcal{P} + \mathcal{P}$.

Proof. We will prove the first item. The second one is an immediate consequence of the first one.

$$c(m) = \psi(m) - \psi(m-1) = \sum_{i=1}^{\pi(m-1)} \pi \, (2m - p_i) - \sum_{i=1}^{\pi(m-1)} \pi \, (2 \, (m-1) - p_i) \geq$$

$$\geq a_1 \sum_{i=1}^{\pi(m-1)} \pi \, (2m) - a_2 \sum_{i=1}^{\pi(m-1)} \pi \, (p_i) + a_3 \, \pi \, (m-1) + a_4 \, 2m \, \pi \, (m-1) -$$

$$-a_5 \sum_{i=1}^{\pi(m-1)} p_i - b_1 \sum_{i=1}^{\pi(m-1)} \pi \, (2 \, (m-1)) + b_2 \sum_{i=1}^{\pi(m-1)} \pi \, (p_i) - b_3 \, \pi \, (m-1) -$$

$$-b_4 \, \pi \, (m-1) \, 2 \, (m-1) + b_5 \sum_{i=1}^{\pi(m-1)} p_i = a_1 \, \pi \, (m-1) \, \pi \, (2m) -$$

$$-a_2 \, \frac{\pi \, (m-1) \, (\pi \, (m-1) + 1)}{2} + a_3 \, \pi \, (m-1) + a_4 \, 2m \, \tau \, (m-1) - a_5 \sum_{i=1}^{\pi(m-1)} p_i -$$

$$-b_1 \, \pi \, (m-1) \, \pi \, (2m-2) + b_2 \, \frac{\pi \, (m-1) \, (\pi \, (m-1) + 1)}{2} - b_3 \, \pi \, (m-1) -$$

$$-b_4 \, \pi \, (m-1) \, 2 \, (m-1) + b_5 \sum_{i=1}^{\pi(m-1)} p_i = (b_2 - a_2) \, \frac{\pi \, (m-1) \, (\pi \, (m-1) + 1)}{2} -$$

$$- (b_1 \, \pi \, (2m-2) - a_1 \, \pi \, (2m)) \, \pi \, (m-1) + (a_3 - b_3) \, \pi \, (m-1) -$$

$$- (b_4 \, (2m-2) - 2a_4 \, m) \, \pi \, (m-1) + (b_5 - a_5) \sum_{i=1}^{\pi(m-1)} p_i.$$

In the first equality we have used that $\pi \, (p_i) = i$ for each $i \in \mathbb{Z}^+$. $\qquad \square$

Theorem 9.5.2. *Let $\mathcal{P} = (p_i)_{i \in \mathbb{Z}^+} \subseteq \mathbb{R}$ be a u.d. strictly increasing sequence and $m \in \mathbb{R} \setminus \mathcal{P}$ be such that $m \geq p_1$, and the distribution function, π, of $\mathcal{P}$ verifies both that $\pi(m) = \pi(m-1)$, and*

$$\pi(2m - p_i) - \pi(2(m-1) - p_i) = \chi_{\mathcal{P}}(2m - p_i) \text{ for each } i \in \{1, ..., \pi(m-1)\}.$$

Suppose that there exist constants $a_1, a_2, a_5, b_1, b_2, b_5 > 0$ and $a_3, a_4, b_3, b_4 \in \mathbb{R}$ such that

$$a_1 \pi(x) - a_2 \pi(y) + a_3 + a_4 x - a_5 y \leq \pi(x - y) \leq b_1 \pi(x) - b_2 \pi(y) + b_3 + b_4 x - b_5 y$$

for all $x, y > 0$, $y \leq \frac{x}{2}$.
    Then

1.

$$0 \leq c(m) = \psi(m) - \psi(m-1) \leq (b_1 \pi(2m) - a_1 \pi(2m - 2)) \pi(m-1) -$$
$$- (b_2 - a_2) \frac{\pi(m-1)(\pi(m-1) + 1)}{2} + (b_3 - a_3) \pi(m-1) +$$
$$+ 2(mb_4 - a_4(m-1)) \pi(m-1) - (b_5 - a_5) \sum_{i=1}^{\pi(m-1)} p_i.$$

2. *Suppose that*

$$(b_2 - a_2) \frac{\pi(m-1) + 1}{2} + (b_5 - a_5) \frac{\sum_{i=1}^{\pi(m-1)} p_i}{\pi(m-1)} =$$
$$= b_1 \pi(2m) - a_1 \pi(2m - 2) + b_3 - a_3 + 2a_4 + 2m(b_4 - a_4).$$

Then $c(m) = 0$, this is, $2m \notin \mathcal{P} + \mathcal{P}$.

Proof. We will prove the first item. The second one is obvious from the first one.

$$c(m) = \psi(m) - \psi(m-1) = \sum_{i=1}^{\pi(m-1)} \pi(2m - p_i) - \sum_{i=1}^{\pi(m-1)} \pi(2(m-1) - p_i) \le$$

$$\le b_1 \sum_{i=1}^{\pi(m-1)} \pi(2m) - b_2 \sum_{i=1}^{\pi(m-1)} \pi(p_i) + b_3 \pi(m-1) + b_4 2m\pi(m-1) -$$

$$-b_5 \sum_{i=1}^{\pi(m-1)} p_i - a_1 \sum_{i=1}^{\pi(m-1)} \pi(2(m-1)) + a_2 \sum_{i=1}^{\pi(m-1)} \pi(p_i) - a_3 \pi(m-1) -$$

$$-a_4 \pi(m-1) 2(m-1) + a_5 \sum_{i=1}^{\pi(m-1)} p_i = b_1 \pi(m-1) \pi(2m) -$$

$$-b_2 \frac{\pi(m-1)(\pi(m-1)+1)}{2} + b_3 \pi(m-1) + b_4 2m\pi(m-1) - b_5 \sum_{i=1}^{\pi(m-1)} p_i -$$

$$-a_1 \pi(m-1) \pi(2m-2) + a_2 \frac{\pi(m-1)(\pi(m-1)+1)}{2} - a_3 \pi(m-1) -$$

$$-a_4 \pi(m-1) 2(m-1) + a_5 \sum_{i=1}^{\pi(m-1)} p_i = -(b_2 - a_2) \frac{\pi(m-1)(\pi(m-1)+1)}{2} +$$

$$+(b_1 \pi(2m) - a_1 \pi(2m-2)) \pi(m-1) + (b_3 - a_3) \pi(m-1) +$$

$$+(2b_4 m - a_4(2m-2)) \pi(m-1) + (a_5 - b_5) \sum_{i=1}^{\pi(m-1)} p_i,$$

where we have also used that $\pi(p_i) = i$ for all $i \in \mathbb{Z}^+$. $\qquad\square$

9.6 Induction on the number of primes in the descomposition of a given number

In this section we will apply induction over the number of primes that constitute a given natural non-zero number in order to obtain some results on Goldbach's Conjecture. Let π be the distribution function of the prime numbers set,

$$\mathbb{P} = \{p_0 = 2,\ p_1 = 3,\ p_2 = 5, ...\}.$$

Let $m \in \mathbb{N}$, $m \geq 3$, $m \notin \mathbb{P}$. Thus $\pi(m) = \pi(m-1)$ and

$$\pi\left(2m - p_i\right) - \pi\left(2\left(m-1\right) - p_i\right) = \chi_{\mathcal{P}}\left(2m - p_i\right) \text{ for every } i \in \{1, ..., \pi\left(m-1\right)\}.$$

Consider

$$\psi(m) = \sum_{i=1}^{\pi(m)} \pi\left(2m - p_i\right)$$

$$\psi\left(m-1\right) = \sum_{i=1}^{\pi(m-1)} \pi\left(2\left(m-1\right) - p_i\right).$$

Suppose that $2m \in \mathbb{P} + \mathbb{P}$, this is, $\psi(m) > \psi(m-1)$. We also have that

$$\psi\left(mp\right) = \sum_{i=1}^{\pi(mp)} \pi\left(2mp - p_i\right)$$

$$\psi\left(mp-1\right) = \sum_{i=1}^{\pi(mp-1)} \pi\left(2\left(mp-1\right) - p_i\right).$$

The key question is if $\psi\left(mp\right) > \psi\left(mp-1\right)$ is true, that is, if $2mp \in \mathbb{P} + \mathbb{P}$.
    Notice that

$$\psi\left(mp\right) = \sum_{i=1}^{\pi(mp-1)} \pi\left(2mp - p_i\right),$$

because $mp \notin \mathbb{P}$.
    Assume that there exist constants $C(p)$, $D(p) > 0$ independent of m such that

a) $\frac{\psi(mp)}{\pi(mp)} \geq C(p) \frac{\psi(m)}{\pi(m)}$.

b) $\frac{\psi(m-1)}{\pi(m-1)} \geq D(p) \frac{\psi(mp-1)}{\pi(mp-1)}$.

c) $C(p) \cdot D(p) \geq 1$.

We claim that in that case we obtain that $\psi\left(mp\right) > \psi\left(mp-1\right)$.

Indeed,

$$\frac{\psi(m\,p)}{\pi(m\,p)} \geq C(p)\,\frac{\psi(m)}{\pi(m)} >$$

$$> C(p)\,\frac{\psi(m-1)}{\pi(m)} = C(p)\,\frac{\psi(m-1)}{\pi(m-1)} \geq$$

$$\geq C(p) \cdot D(p)\,\frac{\psi(m\,p-1)}{\pi(m\,p-1)} = C(p) \cdot D(p)\,\frac{\psi(m\,p-1)}{\pi(m\,p)} \geq \frac{\psi(m\,p-1)}{\pi(m\,p)}.$$

Hence,

$$\frac{\psi(m\,p)}{\pi(m\,p)} > \frac{\psi(m\,p-1)}{\pi(m\,p)},$$

and then

$$\psi(m\,p) > \psi(m\,p-1),$$

whereby our claim is true.

This is the motivation for the next result.

Theorem 9.6.1 (GP). *Suppose that there exist constants a_1, a_2, b_1, $b_2 > 0$ and a_3, $b_3 \in \mathbb{R}$ such that*

$$a_1\,\pi(x) - a_2\,\pi(y) + a_3 \leq \pi(x-y) \leq b_1\,\pi(x) - b_2\,\pi(y) + b_3 \text{ for all } x,\, y \in \mathbb{R},\ y \leq \frac{x}{2}.$$

Let $r \in \mathbb{N}$, $r \geq 3$, $r \notin \mathbb{P}$. Suppose that $2r \in \mathbb{P} + \mathbb{P}$, which is equivalent to, $\psi(r) > \psi(r-1)$. Let $q \in \mathbb{P}$. In addition assume that

1. $C(q) := \inf_{m \in \mathbb{N},\, m \geq 3} \dfrac{a_1\,\pi(2m\,q) - \frac{a_2}{2}\,(\pi(m\,q)+1) + a_3}{b_1\,\pi(2m) - \frac{b_2}{2}\,(\pi(m)+1) + b_3} \in (0,\, +\infty)$.

2. $D(q) := \inf_{m \in \mathbb{N},\, m \geq 3} \dfrac{a_1\,\pi(2m-2) - \frac{a_2}{2}\,(\pi(m-1)+1) + a_3}{b_1\,\pi(2m\,q-2) - \frac{b_2}{2}\,(\pi(m\,q-1)+1) + b_3} \in (0,\, +\infty)$.

3. $C(q) \cdot D(q) \geq 1$.

Then $2r\,q \in \mathbb{P} + \mathbb{P}$.

Proof. Using the bounds for $\pi(x - y)$, x, $y \in \mathbb{R}$, $y \leq \frac{x}{2}$, we obtain:

$$\frac{\psi(r\,q)}{\pi(r\,q)} \leq b_1\,\pi(2r\,q) - \frac{b_2}{2}\,(\pi(r\,q) + 1) + b_3,$$

$$\frac{\psi(r\,q - 1)}{\pi(r\,q - 1)} \leq b_1\,\pi(2r\,q - 2) - \frac{b_2}{2}\,(\pi(r\,q - 1) + 1) + b_3,$$

$$\frac{\psi(r)}{\pi(r)} \leq b_1\,\pi(2r) - \frac{b_2}{2}\,(\pi(r) + 1) + b_3,$$

$$\frac{\psi(r - 1)}{\pi(r - 1)} \leq b_1\,\pi(2r - 2) - \frac{b_2}{2}\,(\pi(r - 1) + 1) + b_3,$$

$$\frac{\psi(r\,q)}{\pi(r\,q)} \geq a_1\,\pi(2r\,q) - \frac{a_2}{2}\,(\pi(r\,q) + 1) + a_3,$$

$$\frac{\psi(r\,q - 1)}{\pi(r\,q - 1)} \geq a_1\,\pi(2r\,q - 2) - \frac{a_2}{2}\,(\pi(r\,q - 1) + 1) + a_3,$$

$$\frac{\psi(r)}{\pi(r)} \geq a_1\,\pi(2r) - \frac{a_2}{2}\,(\pi(r) + 1) + a_3,$$

$$\frac{\psi(r - 1)}{\pi(r - 1)} \geq a_1\,\pi(2r - 2) - \frac{a_2}{2}\,(\pi(r - 1) + 1) + a_3.$$

Let $p \in \mathbb{P}$. The inequality

$$\frac{\psi(r\,p)}{\pi(r\,p)} \geq C(p)\,\frac{\psi(r)}{\pi(r)}$$

is equivalent to

$$a_1\,\pi(2r\,p) - \frac{a_2}{2}\,(\pi(r\,p) + 1) + a_3 \geq C(p)\left(b_1\,\pi(2r) - \frac{b_2}{2}\,(\pi(r) + 1) + b_3\right).$$

The inequality

$$\frac{\psi(r - 1)}{\pi(r - 1)} \geq D(p)\,\frac{\psi(r\,p - 1)}{\pi(r\,p - 1)}$$

is equivalent to

$$a_1\,\pi(2r - 2) - \frac{a_2}{2}\,(\pi(r - 1) + 1) + a_3 \geq$$

$$\geq D(p)\left(b_1\,\pi(2r\,p - 2) - \frac{b_2}{2}\,(\pi(r\,p - 1) + 1) + b_3\right).$$

Hence both inequalities are true by the definitions of $C(p)$ and $D(p)$. $\qquad\square$

Corollary 9.6.2 (GP). *Suppose that there exist constants a_1, a_2, b_1, $b_2 > 0$ and a_3, $b_3 \in \mathbb{R}$ such that*

$$a_1 \pi(x) - a_2 \pi(y) + a_3 \leq \pi(x-y) \leq b_1 \pi(x) - b_2 \pi(y) + b_3 \text{ for all } x, y \in \mathbb{R},\ y \leq \frac{x}{2}.$$

Let $r \in \mathbb{N}$, $r \geq 3$, $r \notin \mathbb{P}$. Suppose that $2r \in \mathbb{P} + \mathbb{P}$, which is equivalent to, $\psi(r) > \psi(r-1)$. In addition assume that

1. $C(p) := \inf_{m \in \mathbb{N},\, m \geq 3} \dfrac{a_1 \pi(2mp) - \frac{a_2}{2}\,(\pi(mp)+1) + a_3}{b_1 \pi(2m) - \frac{b_2}{2}\,(\pi(m)+1) + b_3} \in (0, +\infty)$ for each $p \in \mathbb{P}$.

2. $D(p) := \inf_{m \in \mathbb{N},\, m \geq 3} \dfrac{a_1 \pi(2m-2) - \frac{a_2}{2}\,(\pi(m-1)+1) + a_3}{b_1 \pi(2mp-2) - \frac{b_2}{2}\,(\pi(mp-1)+1) + b_3} \in (0, +\infty)$ for all $p \in \mathbb{P}$.

3. $C(p) \cdot D(p) \geq 1$ for all $p \in \mathbb{P}$.

Then $2r\,p \in \mathbb{P} + \mathbb{P}$ for every $p \in \mathbb{P}$.

Remark 9.6.3. *Obviously, under the assumptions of Corollary 9.6.2 we have that the Goldbach Property for $\mathbb{P}$ is true, using induction over the number of primes whose product form a given natural number greater than 3.*

Namely, we have the following fundamental result.

Corollary 9.6.4. *Assume that there exist constants a_1, a_2, b_1, $b_2 > 0$ and a_3, $b_3 \in \mathbb{R}$ such that*

$$a_1 \pi(x) - a_2 \pi(y) + a_3 \leq \pi(x-y) \leq b_1 \pi(x) - b_2 \pi(y) + b_3 \text{ for all } x, y \in \mathbb{R},\ y \leq \frac{x}{2}.$$

Suppose that

1. $C(p) := \inf_{m \in \mathbb{N},\, m \geq 3} \dfrac{a_1 \pi(2mp) - \frac{a_2}{2}\,(\pi(mp)+1) + a_3}{b_1 \pi(2m) - \frac{b_2}{2}\,(\pi(m)+1) + b_3} \in (0, +\infty)$ for each $p \in \mathbb{P}$.

2. $D(p) := \inf_{m \in \mathbb{N},\, m \geq 3} \dfrac{a_1 \pi(2m-2) - \frac{a_2}{2}\,(\pi(m-1)+1) + a_3}{b_1 \pi(2mp-2) - \frac{b_2}{2}\,(\pi(mp-1)+1) + b_3} \in (0, +\infty)$ for all $p \in \mathbb{P}$.

3. $C(p) \cdot D(p) \geq 1$ for all $p \in \mathbb{P}$.

Then Goldbach Property is true for $\mathbb{P}$ with respect to every $r \in \mathbb{N}$, $r \geq 3$.

Remark 9.6.5. *Notice that Theorem 9.6.1, Corollary 9.6.2 and Corollary 9.6.4 are also true for each u.d. strictly increasing sequence $\mathcal{P} = (p_i)_{i\in\mathbb{Z}^+} \subseteq \mathbb{R}$ such that $\mathcal{P} \subseteq 2\mathbb{Z}$ or $\mathcal{P} \subseteq 2\mathbb{Z}+1$. Furthemore, they are true for every u.d. strictly increasing sequence $\mathcal{P} = (p_i)_{i\in\mathbb{Z}^+} \subseteq \mathbb{R}$ and every $r \in \mathbb{N} \setminus \mathcal{P}$, $r \geq 2$, verifying both that $\pi(r) = \pi(r-1)$ and*

$$\pi\left(2r - p_i\right) - \pi\left(2\left(r-1\right) - p_i\right) = \chi_\mathcal{P}\left(2r - p_i\right) \text{ for each } i \in \{1, ..., \pi\left(r-1\right)\}.$$

Remark 9.6.6. *Under the hypotheses of Theorem 9.6.1 (even for a general set $\mathcal{P}$ as indicated in Remark 9.6.5), a sufficient condition for the bound $D(q)$ to be a positive real number is the following one:*

For every $m \in \mathbb{N}$, $m \geq 3$, there exists $E(m) \in (0, +\infty)$ such that

$$b_1\,\pi\left(2m\,p - 2\right) - \frac{b_2}{2}\left(\pi\left(m\,p - 1\right) + 1\right) + b_3 \leq E(m) \text{ for every } p \in \mathbb{P}.$$

ACKNOWLEDGEMENT

The author would like to express his gratitude to the Office for Education, Culture and Universities of Murcia for its institutional support. This book is dedicated to God, my Fortress; and to Inma, my blessing, my wife.

Bibliography

[1] Bateman, P. T., Diamond H. G., *Analytic Number Theory: An Introductory Course.* World Scientific Pub. Co. Inc., Monographs In Number Theory, 2004.

[2] Chen, J. R., *On the representation of a larger even integer as the sum of a prime and the product of at most two primes*, Sci. Sinica **16** (1973), 157–176.

[3] De La Vallèe Poussin, Ch. J., *Recherches analytiques sur la thèorie des nombres premiers*, Ann. Soc. Sci. Bruxelles **20** (1896), 183–256, 281–297.

[4] Dusart, P., *Autour de la fonction qui compte le nombre de nombres premiers*, Doctoral Thesis, Universitè de Limoges, 1998.

[5] Dusart, P., *Estimates of some functions over primes without the Riemann Hypothesis*, arXiv:1002.0442v1[math.Nt], 2010.

[6] Estermann, T., *On Goldbach's problem: proof that almost all even positive integers are sums of two primes*, Proc. London Math. Soc. **44** (1938), 307–314, doi:10.1112/plms/s2-44.4.307.

[7] Guy, R. K., *Unsolved Problems in Number Theory*, Springer-Verlag, New York, 3rd ed., 2004.

[8] Hadamard, Jacques, *Sur la distribution des zèros de la fonction $\zeta(s)$ et ses consèquences arithmètiques*, Bull. Soc. Math. France **24** (1896), 199–220.

[9] Hardy, G. H., Littlewood, J. E., *Some problems of 'Partitio Numerorum'. III. On the Expression of a Number as a Sum of Primes.* Acta Math. **44** (1923), 1–70.

[10] Helfgott, H. A., *The ternary Goldbach problem* arXiv: 1501.05438v2[mathNT], 2014.

[11] Landau, E., *Handbuch der Lehre von der Verteilung der Primzahlen.* Chelsea, New York, 1953.

[12] López Nicolás, J. A., *Advances in Additive Number Theory* Bol. Soc. Paran. Mat. **44** (2021), ISSN-0037-8712 In Press.

[13] Rosser, J. Barkley, Schoenfeld, L., *Sharper Bounds for the Chebyshev Functions $\theta(x)$ and $\psi(x)$.* , Math. Of Computation **29** Number 129 (1975), 243–269.

[14] Schnirelmann, L. G., *On the additive properties of numbers*, Proceedings of the Don Polytechnic Institute in Novocherkassk **XIV** (1930), 3–27.

[15] Schnirelmann, L. G., *Über additive Eigenschaften von Zahlen*, Mathematische Annalen **107** (1933), 649–690.

[16] Tao, T., *Every odd number greater than 1 is the sum of at most five primes*, arXiv:1201.6656v4 [math.NT], doi.org/10.48550/arXiv.1201.6656, 2012.

[17] Van der Corput, J. G., *Sur l'hypothèse de Goldbach*, Proc. Akad. Wet. Amsterdam **41** (1938), 76–80.

[18] Vinogradov, I, *Representation of an Odd Number as a Sum of Three Primes*, Comptes rendus (Doklady) de l'Acadèmie des Sciences de l'U.R.S.S. **15** (1937a), 169–172.

[19] Vinogradov, I, *Some Theorems Concerning the Theory of Primes*, Recueil Math. **2** (1937b), 179–195.

[20] Vinogradov, I, *The Method of Trigonometrical Sums in the Theory of Numbers*, London: Interscience (1954).

www.ingramcontent.com/pod-product-compliance
Lightning Source LLC
Chambersburg PA
CBHW082150290526
45794CB00008B/3239